Forestry and food security

FAO
FORESTRY
PAPER

90

FOOD
AND
AGRICULTURE
ORGANIZATION
OF THE
UNITED NATIONS
Rome, 1989

M-30

ISBN 92-5-102847-8

Foreword

This report summarizes the current state of understanding regarding the links between forestry and food security. It is the result of a series of investigations begun in 1985 in response to the widely felt concern that excessive deforestation is threatening not only the soil and water base essential for continued food production, but also the present and future availability of the many forest plants and animals that are sources of food.

In April 1985, the 10th Session of the FAO Committee on World Food Security considered a preliminary study of the role of forestry in promoting food security and called for further exploration. During 1986 and 1987, the matter was discussed by the Committee on Forestry and three of the FAO Regional Forestry Commissions; these bodies recommended that an FAO Expert Consultation be convened.

The resulting Consultation, held in Banglore in February 1988 at the invitation of the Government of India, brought together 57 experts from 27 countries and organizations. The scope of the Consultation included all forestry-related activities that have a direct or indirect impact on food production and food security at the local level. Particular attention was paid to the links between social, economic, technical, environmental and institutional issues. Emphasis was also placed on questions of equity, and especially to food security of the poor and other vulnerable groups.

To provide background for the Consultation, a series of papers were commissioned on various aspects of forestry and food security. The present report is a synthesis of the background material and the conclusions and recommendations of the Expert Consultation. The report also points to a number of important areas where information is still lacking, or where differences of view still exist. It is hoped, therefore, that this document will encourage the design of policies and programmes which reflect more fully the impact of the linkages between forestry and food security. At the same time it should stimulate further research aimed at closing the gaps in our understanding.

This publication was funded with inputs from the FAO/SIDA Forests, Trees and People Programme.

C.H. Murray
Assistant Director-General
Forestry Department

Table of Contents

page

Chapter 1   Overview

1.1 Introduction: The concept of food security

Food security is a fundamental problem facing the world today.
Despite substantial increases in food production in many
countries, over 800 million people still suffer from
malnutrition. According to FAO figures, approximately 20 million
people are dying of starvation or related diseases each year.
It is estimated that by the year 2000 up to 70 countries,
including 49 in Africa, will not be able to feed themselves
unless urgent action is taken.

Food security has been defined by the Committee on World Food
Security as the "economic and physical access to food, of all
people, at all times." The concept recognises that the
nutritional well-being of people depends not just on food
production; if that were the case then no-one would go hungry,
since total food production is more than enough to feed the
world's population. Food security is also crucially dependent
on the reliability of production and on people's access to
supplies.   It thereby encompasses questions both of
sustainability and equity.

For many foresters the issue of food security may seem to be a
concern which goes far beyond the domain of their profession.
And yet, in many rural areas forests and farm trees provide
critical support to agricultural production (e.g. maintaining and
improving soil conditions, and maintaining hydrological systems),
they provide food, fodder and fuel, and they provide a means of
earning cash income. Thus, both directly and indirectly forestry
activities may have an impact on people's food security.

Within the community of forestry professionals, food security has
emerged in the last few years as a new focus for forestry
development and planning. While it is recognized that forests
contribute to food security in many ways, these links have seldom
been studied in depth and there have been few attempts to assess
their significance. At the policy and planning level, very
little has been done to incorporate food security as a specific
objective in forestry strategies and programmes.

This report is the result of an Expert Consultation on forestry
and food security sponsored by the FAO Forestry Department (held
in India in 1988). It illuminates some of the links between
forestry and food security, and shows how forestry activities can
and do have an impact on food security. In this report forestry
is defined in a broad sense to include management and use of
trees and shrubs on farms and grazing areas, as well as within
established forest reserves. Drawing on many different sources,
it pieces together a picture of the complex interactions between
people, trees, forests, agriculture and food production. It
looks at negative as well as positive effects of forestry
activities, and it aims to distinguish links between forestry and
food security that are well proven from those that are still

speculative or disputed. Going beyond this, the report also sets
out some initial ideas about how forestry policies and programmes
can be directed to improving food security, especially for the
poor.

The picture this report presents is in no way complete; there
are important information gaps and many of the examples come from
isolated reports that may not be representative. The conclusions
reached should therefore be treated as preliminary. They do,
however, provide a basis for further investigation, and are
intended as a stimulus for the more detailed consideration of
individual cases.

1.2 Putting forestry in perspective

The part played by forestry in food security must be kept in perspective. Forests are just one element within the complex fabric of rural life, and food security depends on a whole range of factors quite apart from forests and forestry activities.

It is clearly wrong, for example, to suggest that forestry can replace agriculture as a food production system to any significant extent. It must also be recognised that forestry initiatives, by themselves, cannot remove the underlying pressures caused by population growth. Neither can they fundamentally alter the social, economic and political factors that create inequalities, and separate the rich from the poor; the hungry from the well-fed.

The premise of this report, however, is that forests and trees do have an important role to play in food security. It is a role that has been ignored in the past, and is currently being eroded as forests in many parts of the world are cleared and the remaining trees on farmland come under increasing pressure. These trends are undermining existing agricultural systems and jeopardising their long-term productivity.

But these trends are not irreversible. Through better management of forests, and by supporting tree growing on farms, the contribution of forestry to food security can be both strengthened and enhanced. Forestry initiatives have the potential for providing a range of benefits - augmenting food production, increasing the sustainability of food supplies, and improving access to food for the landless and poor by providing subsistence products, income and employment.

1.3 The links between forestry and food security

Figure 1.1 highlights some of the important links between forestry and food security and suggests some of the ways forest products and environmental benefits, as well as forestry activities, can have an impact on household food security and individual nutritional well-being. The boxes on the far left represent forest products and benefits on which forestry projects often focus (e.g. shelterbelts and fuelwood production). Moving to the right, the linkages between forestry outputs and household food status are illustrated.

It is clear that many links between forestry and food security are inter-related. To simplify the discussion, however, they can be divided into three main groups: environmental, production, and socio-economic factors.

1.3.1   Environmental Links

Trees and forests influence both their immediate surroundings and the stability of the larger environment, and as a result have several important links to food security. Both at the micro and the macro-level, they help provide the stable environmental conditions on which sustainable food production depends. For many communities in tropical regions forests provide the only means for restoring soil productivity (through systems of forest fallowing). Forest areas also represent the single largest storehouse of genetic diversity, a resource of great importance to future agricultural production.

The effects of trees are most easily seen at the farm level, where they can play an important role in improving the micro-climate, reducing the damage caused by wind, protecting against soil erosion, and restoring soil productivity. At the watershed level, forests can reduce sedimentation and improve water quality; they may also have an effect on water availability downstream, and may assist to some extent in reducing the incidence of floods. All of these factors have a major influence on downstream agriculture. At a regional and global level, forests may also affect climate and rainfall patterns - although the detailed interactions are controversial and still only partly understood.

1.3.2   Production Links

The most direct connection between forestry and food security is the food items produced by trees. Fruits, nuts, leaves, roots and gums are just some of the huge array of edible foods that are obtained from trees and shrubs, either growing naturally in the wild or cultivated on farms and around the home. Forests also provide a habitat for many animals, birds, insects and other forms of wildlife that are hunted and consumed, often as delicacies. While these forest foods rarely provide staples, they do provide important supplements as well as seasonal and emergency substitutes when food supplies dwindle.

Figure 1.1 The links between forestry and household food security

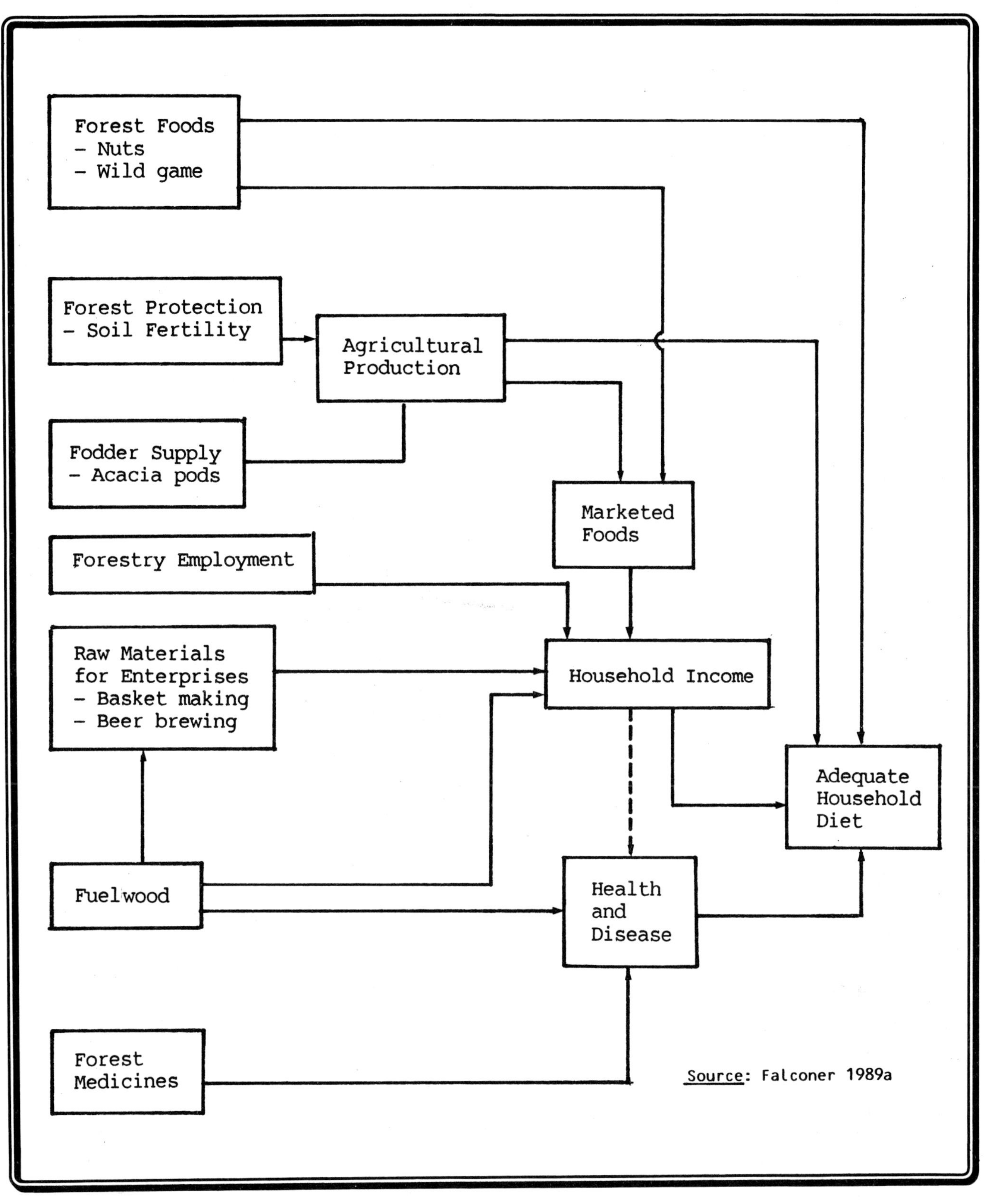

In addition, forests can have an important indirect influence on
food production. By maintaining and improving soil fertility,
trees grown on farms can help sustain crop yields.   In pastoral
production systems, trees and shrubs provide an essential source
of livestock fodder, especially during the dry season. And in
mangrove areas, the forests are a habitat and breeding ground
for many fish, crustacea and other marine animals that support
coastal and off-shore fisheries.

1.3.3   Socio-Economic Links

Food security is fundamentally a social issue. The socio-economic
links between forestry and food security are those that link the
products and "services" of forests to the people who depend on
them. From the point of view of individual households, forests
may affect their food security in several ways. Foods obtained
from trees and forests make an important direct contribution to
family diets, providing a tasty and nutritious supplement to
otherwise bland staple foods. Although the quantities involved
may be small, their nutritional contribution is often critical,
especially at certain times of the year, and during droughts or
other emergency periods when cultivated foods are unavailable.

Even more important for many families is the fact that forests
provide a source of income and employment. Millions of rural
people depend on money earned from gathering, processing and
selling forest products to buy food and other basic necessities.
For the poor, and also for women, these are often one of their
only sources of cash income. Trees grown on the farm are also
used as savings, that can be harvested and sold to meet large or
emergency cash needs.

1.4 Opportunities for action

There is much that can be done by foresters to enhance household
food security. Some of the most obvious opportunities for action
include:

* directing forest management objectives to people's food
  security needs;

* broadening the range of products produced by forests - food
  and other items - and improving their supply to local people
  through new management approaches and access arrangements;

* encouraging tree growing on farms using species and
  management approaches that complement crop and livestock
  production, help protect the environment, provide income to
  farmers, and assist them to spread risks;

* supporting small-scale forest-based enterprises by ensuring
  a sustainable supply of input materials, providing
  managerial and technological assistance, and improving
  access to credit;

* providing market support to help rural people get a better
  price for the forest products they sell, and secure a more
  sustainable livelihood.

While a number of promising approaches of this kind can be
identified, experience in putting them into practice is still
limited. Local circumstances will inevitably play a big part in
determining their relevance and a great deal will depend on local
people's needs, available resources and careful planning.

1.5 Setting the policy framework: new goals and approaches

Forests and farm trees contribute to food security in many rural regions throughout the world. In order to strengthen and develop these contributions, forestry programmes and foresters need to review the goals and devise new approaches for their activities. Existing institutional structures, and the traditional focus of forestry training, research and extension work, are not at all well matched to the task of addressing food security objectives.

Support at the policy level is a prerequisite for change. This means reorganising the specific role of existing forests and of trees in the food security of rural people and their effectiveness in sustaining land-use and food production systems. It will also require support for staff, resources and training. Addressing problems of food security will require a shift in emphasis away from traditional goals of production and protection forestry to gearing forestry activities to meet local people's needs.

Breadfruit - *Artocarpus altalis* a staple food throughout Polynesia

It could mean, for example, upgrading of the status of so-called 'minor forest products' to recognise the extremely important contribution they already make to local incomes and livelihoods, and to exploit the potential for enhancing their production and use. It will involve exploring new approaches to forest management which address issues of access and control of forest resources and which acknowledge the rights of local people to benefit from the forests.

Clearly this will involve putting a lot more effort into understanding local circumstances, and the problems - food security being just one of them - that people face, especially those who are poor. To this end forestry planners will need to build from the considerable traditional knowledge of forest resources that exists in many communities and on methods of managing their local environment.

New types of training will be required for forestry professionals and extension workers to broaden their outlook, and provide them with the skills needed to work more closely with local people. There is a need to bring in other professionals such as nutritionists and social-scientists. Special emphasis must be placed on incorporating the needs and perspectives of women in the planning and implementation of projects.

Much can be gained if forestry services can collaborate more effectively with agriculture departments, and agencies involved in fisheries, livestock and other related professions. Food security crosses over conventional sectoral boundaries and can only be tackled effectively through cooperative endeavours.

More fundamentally, the social, economic and political factors that create and maintain inequalities, and lie behind poverty and hunger, must be recognised. Forestry initiatives cannot change these realities. Even so, there is much that can be done to channel benefits towards poor and disadvantaged groups, provided their needs are properly identified and the necessary commitment exists.

There are many challenges to be faced if forestry is to contribute more effectively to food security. However, there are solid grounds for optimism: forestry philosophy and practice have changed radically over the last two decades, moving away from a narrow traditional view to broader and more people-oriented goals. Incorporating food security concerns can be seen as the logical next step in making forestry more responsive to people's needs, and more relevant to the development process.

Chapter 2 Environmental links between forestry and food security

Sustainable food production depends on a favourable and stable environment. At a local, as well as a regional and global level, trees and forests may have a profound influence on the environment. By protecting the soil from erosion, and stabilising hillsides, exposed coastlines and other fragile areas, they can help preserve the integrity of agricultural land. They may also affect climate and water regimes, both of which are crucial to agriculture.

In some cases the environmental benefits that trees provide are clearly visible. The damage caused by erosion is unmistakable, for example, when steep slopes are cleared of forest. Other environmental influences are much harder to measure. Particularly at the regional and global level it is often difficult to isolate the effects of trees from other factors. A number of controversies remain, and not all of the popular beliefs about the benefits of trees can be backed up by scientific evidence. Care is therefore needed when considering the environmental links between forestry and food security. It is important to distinguish effects that can be clearly demonstrated and relied upon from those that are still speculative, and may depend heavily on local conditions.

2.1 Trees and the microclimate

Interactions between trees and food production are most apparent at the micro-level. Trees, for example, when planted within agricultural areas, have been shown to have a variety of effects on the local microclimate influencing temperature and humidity, moisture availability, and light conditions.

2.1.1   <u>Temperature and Humidity</u>

Tree cover can have a considerable influence in moderating air and soil temperatures, and increasing relative humidity (Lal and Cummings, 1979). Both these effects are generally beneficial to crop growth, a fact that is made use of in many agroforestry systems (Weber and Hoskins, 1983; Vergara and Briones, 1987).

The extent to which these benefits are realised in practice depends on the number of trees involved. An isolated tree planted on farmland can only be expected to have a minor and localised effect. The more the system resembles a closed forest in its canopy structure and tree spacing, the greater the beneficial effect on humidity and temperature.

2.1.2   <u>Shade</u>

The shade cast from trees can have both negative and positive effects. Shade on crops or pasture reduces their photosynthetic activity and beyond a certain point will affect growth rates. Under prolonged and complete shade most annuals and shade-intolerant perennials will die. However, because trees alter temperature and humidity as well, these factors may more than compensate for the reduction in light.

In some cases therefore, varying amounts of shade may benefit different crops. Some types of coffee, for example, are deliberately grown under partial shade. <u>Grevillia robusta</u> has traditionally been used for this purpose in parts of Latin America. A Spanish name for <u>Gliricidia sepium</u>, "madre de cacao" (mother of cacao), indicates its widespread use as shade for cacao plantations. In Sri Lanka, different tree species are used together; some of the best-managed tea estates have a tall shading of either <u>Albizia lebbek</u> or <u>Grevillia robusta</u>, and an intermediate canopy of <u>Gliricidia sepium</u> or <u>Erythrina</u> sp.

Shade may also be very desirable in animal husbandry, particularly in hot climates (Daly, 1984). Though shade will tend to reduce forage production under trees, this is compensated by the protection a tree affords to animals and humans against the hot midday sun. Even single trees are prized in desert or semi-arid situations, such as in the Sahelian and Sudanian zones in Africa, where "every tree is an oasis" (Gorse, 1985).

In this connection, the African species <u>Acacia albida</u> has the unusual feature of being leafless in the rainy season so that there is no shading on crops cultivated beneath it, but full-canopied during the hot, dry season, providing important shade for livestock (Weber and Hoskins, 1983). Manure accumulates where the animals rest, and results in increased fertility not only for the tree, but for crops planted near it (Bonkoungou, 1985).

The overall benefits of shade are not always clear cut. In large intensively-managed monoculture systems shade may prove to be a disadvantage, whereas for less intensive systems, on smallholder properties and on less productive soils shade may provide many advantages (Beer, 1987). Site specific factors are crucial. The benefits of shade depend on the local climate and soils, as well as the particular species involved. For an individual farmer, management requirements of the trees themselves and the marketability of tree products are also important factors.

The trade-offs a farmer has to make in choosing the optimum shade cover are shown clearly in a study in North-east Thailand, where trees are a common feature in most rice paddy fields. The shade they provide was found to be the primary reason for retaining trees on farmland. During the hot, dry season, livestock spend much of their time resting in their shade and grazing in or near shade. Farmers were well aware of the negative effects of too much shade on rice (faster and taller growth - which makes it susceptible to lodging - combined with fewer tillers, less grain and less filled grain). But they felt that the benefits outweighed these costs and controlled shade in many cases by lopping the trees. The species <u>Phyllanthus polythyllus</u> was particularly valued because its sparse foliage does not create too much shade. Its roots help stabilise the crumbly bunds and its branches can be used for bean poles, fences, fuelwood and charcoal (Grandstaff <u>et al</u> 1986).

2.1.3   <u>Moisture Availability</u>

Trees influence the availability of soil moisture in their immediate vicinity. The interception of precipitation by tree foliage influences the amount of moisture reaching the soil. Under a densely crowned tree little or no precipitation may reach the ground in a short, light shower. Only when the tree canopy is fully wetted will most of the precipitation reach the ground. In addition, a tree influences the distribution of moisture reaching the ground. It may come as throughfall (raindrops falling between the leaves), leaf drip or stemflow. The exact pattern depends on the shape of the tree. Understorey plants may find places where stemflow or crowndrip concentrate moisture creating an especially hospitable micro-environment.

Some water is lost through evaporation from the tree canopy. In humid areas, evaporation losses from tree canopies can account for 10 to 30 percent of gross annual rainfall (Vis, 1986). Although a certain amount of evaporation will occur from any surface where water is temporarily held, losses from tree foliage are usually greater than from soil litter or close-to-ground plants, primarily a result of canopy roughness and height (Hamilton and Pearce, 1986).

Uptake of water by tree roots can also have a significant effect on local moisture availability. The impact on crop yields, however, will depend on the extent to which water stress limits crop growth. The drier the environment, the more this is likely to be a problem. The impact will also vary between different species; trees with horizontal surface roots will compete with crops to a much greater extent than deep-rooted species.

2.2 Windbreaks, soil erosion and food crop yields

One of the most widely-recognised benefits of trees on their
immediate environment is their ability to reduce wind speeds.
Farmers in many parts of the world use windbreaks - or more
elaborate multi-species shelterbelts - to protect crops, water
sources, soils and settlements. In addition, windbreaks are
essential first steps for sand dune stabilisation.

There are numerous examples that can be quoted. Tall rows of
Casuarina line thousands of canals and irrigated fields in Egypt.
In Chad and Niger multi-species shelterbelts protect wide
expanses of crop land from desertification. In China, there has
been a massive programme in recent years to establish 'forest
nets' throughout the exposed central plains region.   These
consist of a grid of windbreaks, each 'net' enclosing between 4
and 26 hectares of farmland, depending on the severity of the
wind problem. <u>Paulownia</u> sp. has been the main species used,
because of its deep roots and relatively light shading.

2.2.1   <u>Trees: Barriers Against Erosion</u>

Reducing windspeeds helps substantially in preventing wind
erosion, and the damage it causes (Chepil, 1945). This includes
both damage due to the loss of nutrient-rich topsoil, and damage,
as a result of physical injury to crops and livestock, or the
partial burial of fields. Soils are most susceptible to wind
erosion when they are dry and bare. Thus, overgrazing or any
cropping activity that removes plant cover makes soils more
vulnerable to wind erosion. The hazard increases with the length
of time the soil surface is bare and with the degree of soil
dryness.

Eucalyptus planted to form a windbreak in Tunisia

A well-developed windbreak or shelterbelt can have a considerable
influence in reducing wind velocity at the soil surface. Where
the barrier is at right angles to the wind direction, this effect
has been found to extend up to 5-10 times the height of the
barrier to the windward side, and 30 to 35 times the height to
the leeward, or down-wind side. Small reductions in wind speed
can have a significant impact on soil erosion, in part because
the drying rate of the soils is reduced after rain storms.

A windbreak using a mix of species provides an efficient
semi-permeable barrier to wind over its full height. This
produces a diverse shape to the windbreak as well as ensuring a
long life of the windbreak (by mixing species with varied growth
rates). In addition, a mixture of species provides risk
protection against unexpected attack from diseases or insects
that could destroy single species stands. Trees scattered
throughout the fields such as the <u>Acacia albida</u> parkland savanna
of West Africa can have the effect of breaking up wind patterns
resulting in impacts similar to more formal windbreaks and
shelterbelts.

2.2.2   <u>Other Benefits of Windbreaks</u>

As well as reducing wind erosion, windbreaks and shelterbelts can
benefit agriculture in a variety of other ways:

* windbreaks and shelterbelts help prevent mechanical damage
  caused by high winds (Guyot, 1986). Winds in excess of 8
  metres per second, for example, can break off twigs and
  small limbs in orchard crops. Such losses of photosynthetic
  surface reduce production, and can adversely affect
  flowering and fruiting the following year. Flowers on crops
  are particularly susceptible to high winds, and fruits may
  also be damaged or dislodged. With cereal crops, stem
  breakage or flattening (lodging) is an increasing hazard as
  the crop matures;

* shelter provided by windbreaks helps reduce the rate of
  water loss from crops through evapotranspiration; this can
  extend to as much as 30 times the height of the tree barrier
  (Konstantinov and Struzer, 1965);

* reductions in wind velocity can prevent adverse
  physiological changes in crops - such as the reductions in
  leaf area and photosynthetic rate that are characteristic
  of some crops when exposed to high winds (Whitehead, 1965);

* trees and shelterbelts protect livestock, particularly young
  animals, against the damaging effects of both cold and hot
  winds;

* windbreaks provide an essential element for dune
  stabilization;

* trees planted along coastlines can protect crops from salt
  spray and thus allow farming to be extended closer to the
  shore. The trees selected for such "salt-breaks" must have
  some degree of salt tolerance, for they will concentrate
  salt under their crowns. Species that have been used
  successfully include <u>Casuarina equisetifolia</u>, <u>Casuarina
  glauca</u>, <u>Pinus pinaster</u>, <u>Pinus radiata</u>, and <u>Cupressus
  macrocarpa</u>;

* windbreaks can reduce evaporative losses from ponds,
  irrigation canals and other water bodies, thus making more
  water available to food production;

* by reducing wind velocities windbreaks can help improve
  insect pollination of crops. This is particularly important
  in fruit orchards (Caborn 1965). Beekeepers also find wind
  protection for their hives to be desirable in areas with
  strong, cold or hot winds;

* windbreaks may benefit crop yields by reducing the incidence
  and severity of pest damage. Studies of the Colorado
  beetle, for instance, showed large reductions in populations
  of eggs and larvae near to windbreaks, and higher predator
  densities closer to the trees (Karg, 1976). The effects are
  not uniform, however, since windbreaks can harbour harmful
  pest species as well as pest predators (Janzen, 1976).
  Trees have been traditionally thought to encourage tse-tse
  flies, though this view is not universally accepted. Kenyan
  and Tanzanian experience suggests that windbreaks need not
  shelter tse-tse flies if the under-storey is relatively
  open, the over-storey is high, and the ground surface is
  kept weeded;

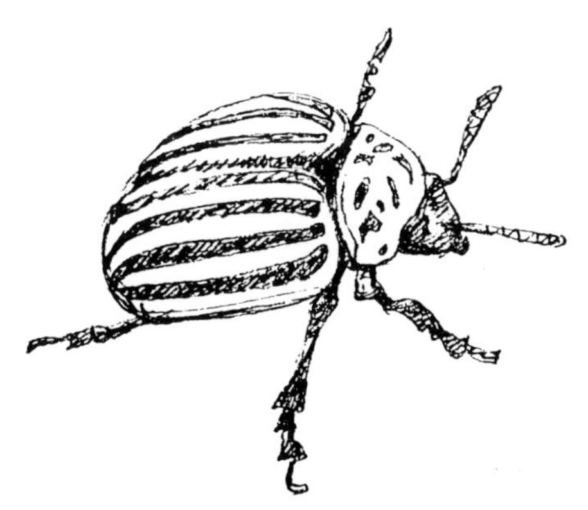

Colorado beetle

Tse-tse fly

* windbreaks can help prevent the spread of plant diseases by
  inhibiting the aerial dispersal of disease spores. This
  effect may be offset, however, by the more rapid development
  of disease spores near to windbreaks resulting from higher
  relative humidity (Guyot, 1986).

As well as reducing windspeeds, windbreaks and shelterbelts provide a range of direct benefits from the fodder, fruit, wood and other products they supply. Even in the harsh desert environment of Yemen, a two row windbreak of _Conocarpus lancifolius_ yielded 350 m3 of wood per kilometre every 20 years, which was more than enough to offset the establishment costs, without considering the additional agricultural benefits (Costen, 1976). In the Majjia Valley in Niger, pollarding of windbreaks every four years is estimated to bring local residents US$ 800 worth of construction poles and wood per kilometre of windbreak (USAID, 1987). Several books and manuals on the subject of windbreak design are available (see Guyot, 1986; Bhimaya, 1976; Weber, 1986).

2.2.3   _Effects of Windbreaks on Crop Yields_

The effects of windbreaks on crop yields is illustrated in Figure 2.1. Close to the windbreak, yields are reduced because of shading, root competition, and the physical space taken up by the trees. Moving further away the benefits become more apparent, until at a certain distance they start to drop off again as the influence of the trees diminishes.

Some of the most dramatic yield increases have been reported in China, where the hot, dry summer winds represent a major limitation on agricultural production. In Hetian Prefecture, where 110,000 hectares of land were planted with _Paulownia_ sp. windbreaks in the early 1980s using the 'forest net' system, increases in grain yield of 60% have been recorded, along with a 70% increase in natural silk production, and a 300% increase in cotton output (Wang Shiji, 1988).

Significant increases have also been reported in a number countries with Mediterranean type climates. According to one survey covering Argentina, Bulgaria, California, Egypt, Israel, Italy, Saudi Arabia and Tunisia, well-designed windbreaks have given a net increase in crop yields of between 80 and 200 percent (Jensen, 1984). Similar increases have been reported in studies in the Antilles on vegetable yields (Guyot, 1986).

In the Sahel, while statistically valid results are not yet available, initial trials with millet and sorghum suggest that in fields protected by windbreaks yields can be as much as 23 percent higher than in unprotected fields (Bognetteau-Verlinden, 1980). In a year with poor rainfall, even relatively small differences in crop yields can be of major significance to local people.

However, the overall effects of windbreaks on crop yields varies considerably. In some cases yields are increased significantly; in others the competition for water and light and the loss of planting area have been found to be detrimental to crop yields. As a general rule, where land is exposed to high winds for most of the year, or where soil erosion is a particular problem, the case for windbreaks will usually be strong. Where these

conditions do not prevail, the advantages of windbreaks may be
less clear. As well as the direct costs for labour and planting
material, windbreaks will take some land out of crop production,
and will compete for water, light and nutrients. Therefore
windbreak products such as fodder, fuel, and foods, increased
crops yields, and soil improvements must be sufficient to cover
these costs. In many cases, for the farmer the negative effect
on crop yields may be more than made up for by the wood and other
products provided by the windbreak itself and having this dual
production system may reduce risks should one of the systems
fail.

Figure 2.1   Effect on Output of a Field Receiving Protection by a Windbreak

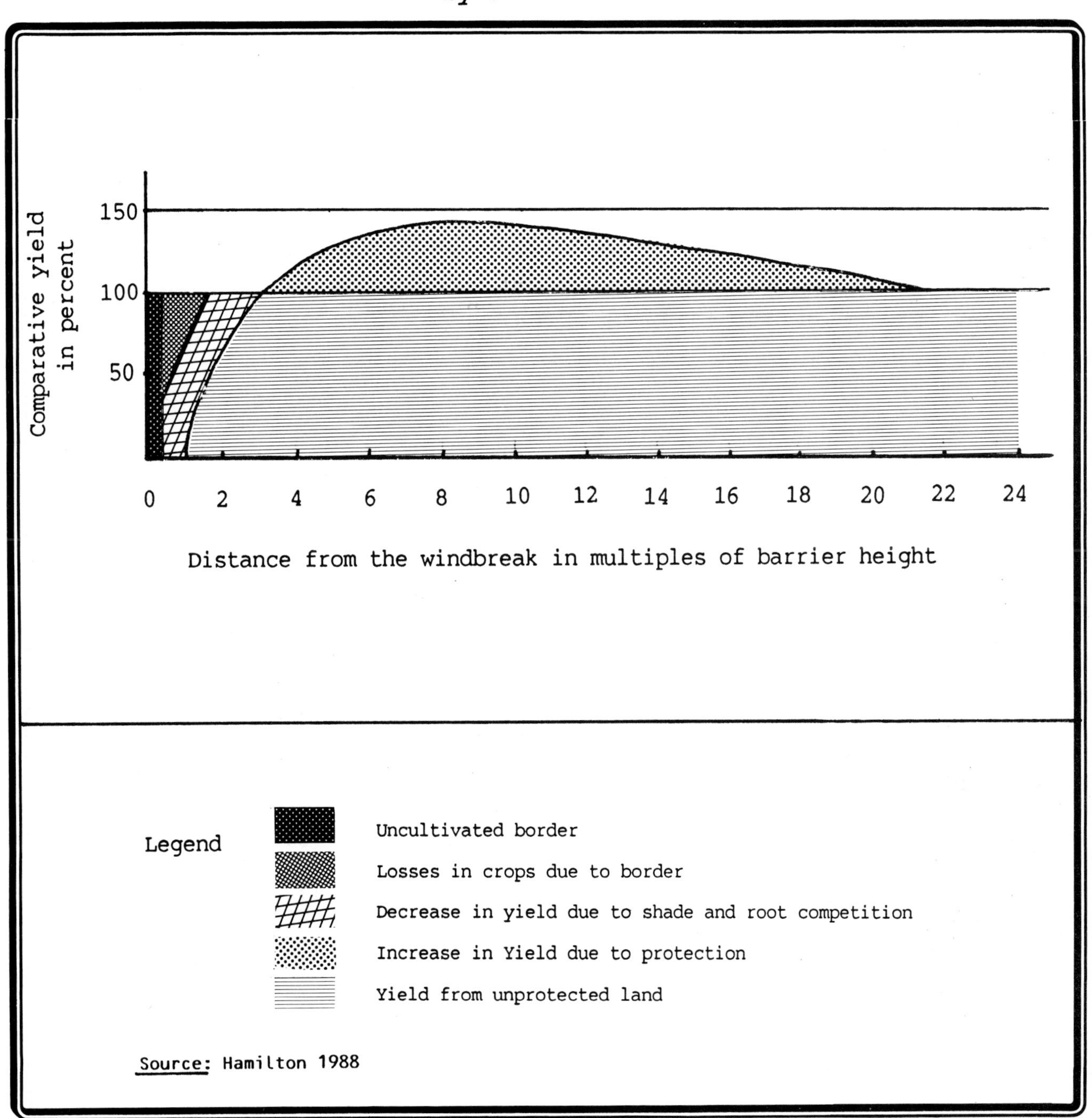

Source: Hamilton 1988

2.3 Tree's role in preventing water erosion

Soil erosion caused by water is a serious problem for agricultural production in many regions of the tropics and subtropics. It strips the most fertile top layers of soil, and can destroy crops themselves by flooding. Forests (and trees) can provide protection against some types of water-induced soil erosion. Surface erosion caused by water in undisturbed forests is generally less than that under other types of land use (Hamilton, 1983). Removing forest, and leaving the soil exposed, can thus have a radical effect in increasing the rate of soil erosion.

Contrary to what is often assumed, it is not the tall tree canopy that gives the most protection to the soil, but the ground cover and litter layer beneath it (Hamilton, 1986). If the ground below is bare, large water droplets falling from a tall tree canopy may actually cause splash erosion and initiate more sheetwash than rain falling on bare soil in the open (Lembaga Ekologi, 1980). Often, therefore, it is not the cutting of trees that leads to surface erosion, but the disturbance to the understorey and leaf litter, and baring of the soil, associated with tree cutting.

TABLE 2.1   Erosion under various tropical moist forest and tree crop systems (tons/ha/year)

	Minimal	Median	Maximal
Multistoried tree gardens	0.01	0.06	0.14
Natural forests	0.03	0.30	6.16
Shifting cultivation, during fallow period	0.05	0.15	7.40
Forest plantations, undisturbed	0.02	0.58	6.20
Tree crops with cover crop/mulch	0.10	0.75	5.60
Shifting cultivation, during cropping period	0.40	2.78	70.05
Taungya cultivation	0.63	5.23	17.37
Tree crops, clean-weeded	1.20	47.60	182.90
Forest plantations, burned and litter removed	5.92	53.40	104.80

<u>Source:</u> Wiersum 1984

The importance of ground cover in protecting against surface erosion has been demonstrated in studies of different forest and tree crop systems, the results of which are summarised in Table 2.1. Forest and tree crop plantations in which the ground cover had been removed were found to be far more susceptible to erosion than those in which it was retained. Similarly, taungya systems (systems where food crops are grown between young plantation trees) were more prone to erosion when the ground between trees was weeded than when a cover crop or mulch layer was maintained.

As slopes increase, in both steepness and length, the dangers of erosion become greater. A variety of soil conservation techniques can be used to reduce erosion. When combined with physical measures, such as terracing, planting of trees and shrubs can help considerably in binding the soil and preventing water erosion.

Such techniques are used in many traditional agroforestry systems (Vergara and Briones, 1987, Nair, 1984a). For example, in Amazonian Ecuador contour strips of <u>Inga edulis</u> (a leguminous fuelwood tree) are used in a cassava swidden system (Bishop, 1983). After the cassava has been harvested, a perennial legume ground cover of <u>Desmodium</u> is planted and grazed by sheep. This represents a combined tree/ground cover/livestock system that, if properly managed, maintains good soil stability and rapidly improves the soil during the fallow period.

It is important to recognise, however, that planting trees does not guarantee effective soil erosion control. The design and management of such systems are crucial. Simply putting trees into a cropping or grazing system, and even complete reforestation, will not eliminate surface erosion.

Forestry activities such as plantations can also increase the possibilities of water-induced soil erosion. For example, serious erosion problems have been reported under teak plantations in Trinidad. This was due to the lack of understorey vegetation and surface litter (Bell, 1973). For the same reason, introducing trees as part of agroforestry systems will not cure the erosion problem if the soil between the trees is bare for most of the year (Hamilton, 1986).

Nevertheless, it is widely recognised that combining trees with other soil conservation techniques can greatly extend the possibilities for sustainable crop cultivation on sloping land. At some stage, however, even the best soil conservation techniques come up against economic or physical barriers which make them impractical. On these sites, there is a very strong case for maintaining, or restoring, undisturbed forest cover.

2.4 Protection afforded by forests in critical or hazardous areas

In environmentally sensitive areas, forests can play an important indirect role in enhancing food security by protecting cropland and grazing areas from natural hazards such as landslides and coastal erosion. In such hazardous areas the removal of forests can seriously jeopardise agricultural production.

2.4.1   Unstable Slopes

The effect of landslides on downslope agricultural production and human settlements can be disastrous. In addition to the immediate physical destruction, the dumping of large amounts of sediment into streams and rivers also affects the quality of water and survival of fisheries downstream. The loss of food supplies combined with increased incidence of diseases (a result of poor water quality) can have a major negative impact on household food security.

Deep-seated slides need to be considered separately from shallow ones. Deep-seated slides are induced largely as a result of the nature of geological material and are little influenced by the presence or absence of trees (Megahan and King, 1985). They can occur on relatively gentle slopes as well as on steep slopes. These sites are risky even for timber production, since logging can have a triggering and destabilising effect. Areas prone to this type of erosion should be left undisturbed, or harvested manually.

Shallow slides or slips are greatly influenced by vegetation.
Tree roots can increase substantially the stability of slip-prone
slopes. Studies in New Zealand found that tree roots provide up
to 80 percent of the soil shear strength under saturated soil
conditions (O'Loughlin and Watson, 1981). In this respect, trees
are much more effective than either crops or grasses. Removing
them can increase the frequency of slides by as much as seven
times (Swanson et al, 1981). In these types of areas forest
management (as well as some agroforestry techniques) can help
protect against slides. Where trees are to be harvested for wood,
coppicing species are most appropriate as their root systems
remain alive and maintain shear strength.

2.4.2   Coastal Protection

Trees have an important function in some coastal areas in
protecting coastlines against wave damage during storms. They
can also help dampen the effects of extreme tidal surges, thereby
protecting inland areas from inundation and physical damage. Thus
trees may help sustain agricultural production in coastal
regions.

Mangrove forests have a particularly significant role in this
context, providing agricultural lands and human settlements with
protection on exposed coastlines (Hamilton and Snedaker, 1984).
While mangrove forests cannot prevent tidal waves and other
natural disasters, they can help in mitigating their effects.
In the case of the Sunderbans forest in Bangladesh, for example,
where tidal surges have resulted in major loss of life and
property damage, the effects would undoubtedly be worse if the
mangroves were to be removed. In addition mangrove regions
provide a protective habitat for many species of coastal fish and
crustacea, thus protecting and sustaining an important food
source for coastal communities.

A mangrove village in Thailand

2.4.3   <u>Riparian Forests</u>

Forests bordering lakes and streams, known technically as
riparian forests, are also valuable for maintaining environmental
stability. They act as vegetative buffers that help prevent
sediment reaching rivers and streams, as well as providing an
important habitat for wildlife. By trapping agricultural
chemicals and pesticides from overland water flow a forest buffer
zone can also contribute to downstream water quality. Riparian
buffer zones are also important for the maintenance of some fish
species; they help maintain stable water temperatures and prevent
sedimentation both of which are important for maintaining fish
populations.

Trees, in addition, can help stabilise river banks and prevent
erosion damage and flooding during storms. In streamside areas
that are experiencing problems, establishing a forest strip can
be an important supplement to structural measures.

The effects of trees along streamsides is not always positive,
however. Trees can use large volumes of water, and in arid or
semi-arid areas this may reduce downstream water yields,
particularly during the dry season (Hough, 1986).

2.4.4   <u>Areas Prone to Salinisation</u>

Soil salinisation and related phenomena are among the most
serious problems threatening land productivity in arid and semi-
arid regions, especially in irrigated areas. In some instances
trees can help mitigate against increasing soil salinisation. In
addition, in some cases forest removal will result in increased
salinisation. Trees often absorb more water than agricultural
crops, forest clearance can thus lead to a rise in the watertable
level (Hamilton, 1983). Where groundwater levels come to within
one metre of the soil surface, capillary action can draw water
to the surface where salts may then be concentrated by surface
evaporation (Hughes, 1984). Seepage of water may also result in
downslope salinisation, and if salts enter water courses this can
impair fish life and make the water unfit for irrigation uses.

Identifying problem areas prior to forest clearance is essential
if salinisation is to be avoided. Where salinisation has already
taken place, the introduction of trees can often play a useful
role in the rehabilitation of the land for agriculture.

2.4.5 Dune Stabilisation

Moving sand dunes pose a major threat to agriculture in many
countries. Combined with other measures, including various
mechanical fixation approaches, trees play an important part in
stabilising dunes and preventing the damage they cause (Weber,
1986; FAO, 1985). Maintaining as complete a vegetative cover as
possible, and reducing wind velocities using windbreaks are often
the best ways of preventing soil movement.

Once dune erosion has begun, however, the first step is to
determine why natural vegetation is not re-colonising the area
(Weber, 1986). If animals or fire are causing the problem, then
tree planting or revegetation alone will not suffice. Fencing
or firebreaks may be a prior requirement, or may in themselves
be sufficient to allow natural regeneration.

2.5 Forests and water supply

The impact of forests on groundwater supply and stream flow is an extremely important issue, especially with regards to food production and food security; but this subject is surrounded by a great deal of myth and misunderstanding. Hydrological systems are complex. While forests can play a variety of useful roles, the assumption that forests are always good for water supplies is a serious over-simplification. Much depends on soil depth, land-use practices, and a range of other factors.

2.5.1   <u>The Effects of Forest Cover on Stream Flow and Groundwater Levels</u>

There is a common notion that more water is released into streams from forests than other kinds of land area, and that the removal of forests results in less water being available downstream. It is also widely held that forest removal lowers the watertable and thereby adversely affects water availability from wells and springs.

While true in some instances, these assumptions are not universally valid. It is often very difficult to predict the exact impact of deforestation, or reforestation, on a particular watershed without concrete evidence from similar circumstances.

Forests use more soil water in converting sunlight to biomass than most other forms of vegetation. As a result, when forests are partially or completely removed, water consumption will drop and the total annual water yield in streams from the area can be expected to increase (Bruijnzeel, 1986; Hamilton, 1983). Increases in stream flow are greatest in the period immediately following forest removal (Bosch and Hewlett, 1982). Water levels are reduced if forest regrowth is vigorous and in some cases water "consumption" can even exceed that of the original forest (Langford, 1976).

The establishment of a forest plantation will tend to reduce streamflow. The faster the rate of growth of the trees, the more pronounced this effect will be. One study in India reported a decrease in water yield of 28 percent following establishment of Eucalypt plantations (Mathur <u>et al</u>, 1976). Although they have become the centre of current controversy about the undesirable effects of tree plantations on water supplies, Eucalypts are not unique in their demand for water. Any tree that is well adapted to the particular site and produces a large amount of biomass – whether it is <u>Eucalyptus</u>, <u>Pinus</u>, <u>Leucaena</u>, or any other species – will also consume large amounts of water.

While these are the general trends, it is important to recognise that variations and exceptions do occur. Streamflow from forests depends on the depth of the soil as this influences the water uptake by trees. On the one hand, where soils are deep, trees with deep roots can extract water that is unavailable to other plants. Uptake and transpiration therefore tend to be considerably greater than for other types of vegetation. On the other hand, on shallow soils, uptake by trees may be comparable to that of vigorous grasses and increases in total annual streamflow will therefore not be great following forest removal.

The effect of tree cover on groundwater levels is similar to that on streamflow. Where on-site groundwater levels have been measured, in the majority of cases the level has been found to rise following forest removal, and fall if an open area is planted to trees (Boughton, 1970; Holmes and Wronski, 1982).

When considering the link between deforestation and groundwater levels, it is important to distinguish the effect of tree cutting from what happens to the land after it has been cleared. If logging or agricultural practices are poor, the resulting compaction of the soil surface may have a significant effect in reducing water infiltration rates resulting in an overall decrease in groundwater levels - and hence lower levels in wells and less reliable springs. Although deforestation is usually blamed in such cases for the reduced water availability, this is often an over-simplification. In reality it is the way the land is treated after tree cutting that causes the problems, not tree cutting itself.

On land that is already badly compacted, tree planting may help to break up the soil structure and consequently increase infiltration rates. Although it has not been confirmed experimentally, this improved water recharge may in some cases be sufficient to compensate for the increased evapotranspiration from the trees. Again, therefore, the overall effect may be contrary to the general trend; tree planting may lead to a rise in groundwater levels.

This brief review of information on the relationship between forests and water supply highlights the complexities involved in managing forests for water supply. There can be little doubt, however, that the resulting impact - whether increasing or decreasing water supplies can have important repercussions for food production.

2.5.2   <u>Forests and Stormflows</u>

From the point of view of agriculture, the variation of flow in rivers and streams is often as or more important as the total amount available for the entire year. Excessive stormflows, and the floods they cause, can have a disastrous impact on downstream agriculture. Fisheries may also be affected because floods usually carry large amounts of sediment, which disrupt the habitat and life cycle of aquatic species. Increased variability in stormflows, therefore, can increase the risk involved in food production.

When forest land is converted to agriculture, the effect on stormflows depends on the type of agricultural practices introduced. Any cropping or grazing methods that cause soil compaction will tend to reduce infiltration rates and result in more water entering streams than would be the case under forests, thus increasing the likelihood of floods. Some soil and land management practices, in contrast, are less likely to create problems. The centuries-old rice terraces found on steep slopes in parts of Java, Bali, Cebu, Nepal and elsewhere testify to the effectiveness of traditional water management techniques. Conversion to agriculture, therefore, does not necessarily increase the susceptibility to flooding.

It is often claimed that forests prevent floods and that the removal of forests helps cause floods. While there is some truth in this, the evidence suggests that the impact of forests is mainly localised and restricted primarily to the more frequent, short-duration storms, rather than major storm events. While studies do demonstrate that deforestation usually results in greater stormflow volumes and higher peakflows in streams in the immediate vicinity, these effects are negligible over large river basins (Reinhart et al, 1963; Douglass, 1983). There is no simple cause-effect relationship between forest cutting in the headwaters and floods in the lower basin (Hewlett, 1982).

The value of forests in reducing flooding is likely to be greatest on deeper soils. By breaking up the soil and improving infiltration, trees help increase water storage capacity. At a certain point, however, any soil will become saturated. Beyond that, trees cannot prevent rainfall from running off into streams.

When unusually heavy storms occur, floods are liable to happen whatever vegetation is present. Catastrophic floods on large rivers are not caused by deforestation but by too much rain falling in a given period, or by the rapid melting of snow. Even large-scale reforestation of upland areas is unlikely to have a significant impact on the frequency of these kinds of flood events. They will happen whether trees are present or not.

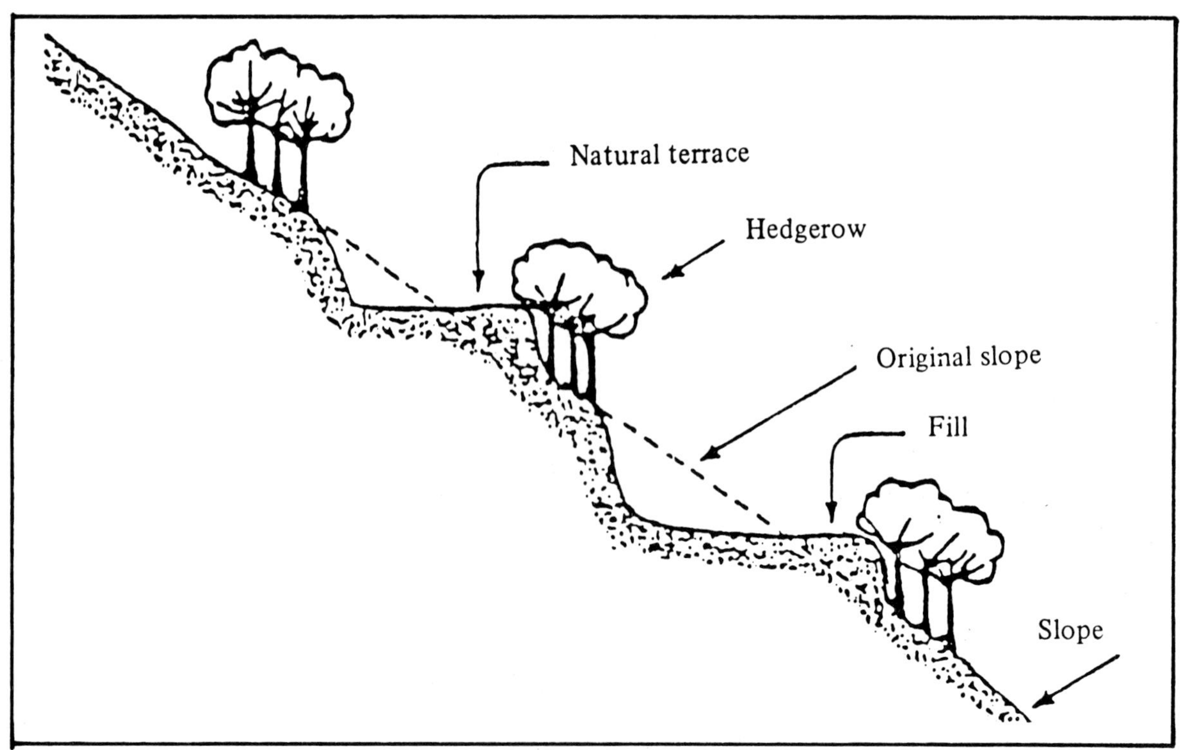

2.5.3   <u>Low Flows</u>

Water shortages caused by reduced dry season flow is also a major
threat in many agricultural areas. It has been suggested that
forests and forest soils may be able to play a beneficial role
by acting as a 'sponge', soaking up water in the rainy season and
then releasing it in the dry season. Tree clearing is supposed
to eliminate this sponge effect and decrease dry season flow
(Spears, 1982).

In practice, there is little scientific evidence to substantiate
this. Most of the experiments that have been carried out have
indicated that cutting trees increases dry season flows in
streams from the treated area, and that planting of trees
decreases flows (Hamilton, 1983). For example, in Northern
Queensland, Australia, a stream draining the area that used to
dry up periodically prior to the rainy season, remained perennial
following logging (Gilmour, 1971). In Fiji, the planting of
<u>Pinus radiata</u> in a dry grassland zone resulted in a 65 percent
reduction in dry season streamflow (Kammer and Raj, 1979).

Making generalisations based on isolated examples is clearly
risky, as a number of factors are at work. Although trees may
help increase infiltration during the rainy season, during the
dry season they draw water from lower depths than might otherwise
contribute to low season river flow or a higher watertable.
Which of these effects is most significant will depend on the
local site conditions.

2.6 Forests, sediment and water quality

Sustainable food production depends on water quality as well as quantity. High levels of sediment and dissolved minerals in rivers and streams can have a variety of negative effects on downstream agriculture and fisheries as well as on peoples' nutritional well-being. By helping to preserve the quality of water supplies, trees and forests play an important role in maintaining food security. Water quality is directly linked to the prevalence of human diseases, especially gastro-intestinal disorders which directly affect people's ability to absorb foods (and thus their nutritional status). It is important to note that the issue of food security includes problems relating to people's ability to use food which is available to them.

From the point of view of nutrient cycling, undisturbed forests are the most efficient of all land use systems (Bormann and Likens, 1981). These systems are even capable of removing and immobilising some potentially harmful pollutants arriving from rain deposition (Sicamma and Smith, 1978). Removal of the forest, partially or completely, breaks this tight chemical cycling and releases minerals and nutrients into the drainage water. This has been shown in studies in Nigeria (Kang and Lal, 1981), Indonesia (Bruijnzeel, 1983) and a number of other countries. Besides causing a loss of nutrients from the site this may also result in an unwelcome addition of nutrients to irrigation water, which can contribute to the eutrophication of water bodies.

The effects of increased sediment levels are usually serious. While modest amounts of sediment may benefit food production in some circumstances - floodplain farmers in Bangladesh, for example, depend on periodic inundation and deposition of nutrient-rich sediment to maintain soil fertility - much more often the effects of sediment are harmful and costly. It can bury crops in floodplains, clog the gills of fish, damage marine fisheries by destabilising mangroves and blanketing seagrass beds and coral reefs, impair the quality of drinking water leading to increased prevalence of diseases, reduce irrigation reservoir capacity, block irrigation canals, and aggravate flooding by filling up flood control dams.

Vegetation cover is not the only factor that influences sediment yield from a given watershed or basin. It is also affected by climate (particularly rainfall), geology, and soils as well as forest fires (Pearce, 1986). If soils are unstable, and rainstorms severe, heavy sedimentation can occur even from watersheds that are entirely forested.

Under given conditions, however, forests do have an important impact in reducing sediment yield from watersheds. In studies in Indonesia, it was found that the sediment yield from areas that had been reforested was only a third of that from an agricultural watershed (Hardjono, 1980). Introducing trees into grazing or cropping land in a well-managed agroforestry system can also have valuable impact on reducing sediment yield (Hamilton, 1983).

The benefits of tree planting for reducing sediment levels may
take years to materialize depending on the transport and storage
mechanisms at work. Because sediment can be trapped and stored
by vegetation and other physical barriers, eroded soil does not
always appear in rivers immediately. There is usually a time
lag, and the larger the watershed and the greater the
opportunities for sediment storage, the longer this lag tends to
be. Changes in erosion rates resulting from altered land-use
practices may therefore take a considerable time before they are
reflected in sediment loads in rivers.

In the case of large watersheds, trapped sediment will continue
to flush out for decades. Reforestation of upland areas may
therefore have little effect in the short term. Reservoirs will
continue to fill with sediment even if the entire above-reservoir
watersheds have been planted in forest. This time lag means that
actions to prevent sedimentation need to be incorporated into the
startup of development projects which are likely to cause (e.g.
road building and logging) or be affected by increased
sedimentation (e.g. dams).

Where sedimentation is a problem, it is important to identify the
precise sources. In a particular watershed, ninety percent of
the problem may be coming from five percent of the land. On
steep terrain, a major source of sediment is often roads and
forest logging activities. Where roads are constructed across
or along stream channels, they put substantial amounts of soil
directly into the waters during the construction phase. If they
are badly located or designed, or improperly maintained, they may
continue to be a serious sediment source for years to come.

2.7 Forests and the global climate

In the long term, one of the most important ways in which forests may influence food production is through their effect on the global climate: altering rainfall patterns, world temperatures, and seasonal climate variations. Cutting of tropical forests has been implicated as being one of the contributory factors behind the gradual increase in the levels of carbon dioxide and certain other trace gases in the atmosphere. The impact this has on the global heat balance, the 'greenhouse effect', has become a cause of widespread concern (Swaminathan, 1986). The two most significant ways in which forests are thought to influence the global climate are through reflected heat from forest areas, and on the level of carbon dioxide in the atmosphere.

While the fact that the carbon dioxide levels are increasing is now widely accepted, the impact of this on the global climate is extremely difficult to estimate, and remains the subject of considerable controversy. Short-term effects may be different from those in the long term, and the effects will almost certainly vary between regions (Henderson-Sellers and Gornitz, 1984).

Agriculture will also be seriously affected if the observed increase in sea level continues, especially in low lying coastal areas. Bangladesh, for example, could lose 10 percent of its land area. Many coastal wetlands and mangrove areas would also be destroyed, with serious consequences for the world's fisheries.

2.7.1 <u>The Albedo Effect</u>

Closed-canopy forests absorb more solar radiation than any other vegetation type and reflect less heat back into the atmosphere. The fraction of reflected radiation is known as "albedo". In recent years there have been many warnings that large scale forest removal may result in an increase in albedo (Hamilton, 1976; Chambers, 1980).

The overall effect of massive deforestation is not easy to predict because, along with increasing the albedo, deforestation may alter other variables that may produce countervailing effects. Two of the most comprehensive 'Global Circulation Model' studies have produced almost exactly opposite predictions about the effects of tropical forest removal. One suggests a slight warming and an increase in rainfall (Lettau <u>et al</u>, 1979). The other predicts a slight cooling in the equatorial region, and an 11 percent reduction in rainfall in the tropics (Potter <u>et al</u>, 1975). A more recent study examining the impact of deforestation of the Amazon rainforest asserts that although radical alteration in forest cover would increase local albedo, there would be no major impact on regional or global climate (Henderson-Sellers and Gornitz, 1984).

With the inherent complexities of climatic systems, all of these
models suffer from problems. Until there are more reliable input
data and better models, there is unlikely to be a clear verdict
on the impact of increased albedo from forest removal on the
global climate.

2.7.2   <u>Carbon Dioxide</u>

Many doubts also exist with regard to the effect of forest
clearance on atmospheric carbon dioxide levels (Woodwell <u>et al</u>,
1978; Hampicke, 1979). Although the cutting and burning of
trees does release carbon dioxide, it is not the only factor
involved; the burning of fossil fuels and cement manufacture are
reckoned to be more important contributors to the increased
carbon dioxide burden.

The global carbon cycle is still only partly understood and there
is considerable scientific disagreement about how much
contribution forest disappearance is really making to increases
in carbon dioxide levels. The fact, for example, that the forest
area in the northern temperate zone has been increasing over the
past five decades may partially offset the large forest loss
occurring in the tropics (Sedjo and Clawson, 1984).

While allowing for factors such as forest burning, the carbon
fixation of forest regrowth, and the effect of carbon levels on
plant photosynthesis levels most models agree that there will be
a considerable net transfer of carbon to the atmosphere from
forest clearance and burning in the tropics. One estimate puts
the total contribution at between 1 and 4.5 billion tons per
year, plus an additional 2 billion tons from oxidation of exposed
organic matter in the soil. This is clearly a substantial
amount, given that fossil fuel burning is currently estimated to
supply around 5 billion tons (Myers, 1980). However, as was noted
above the whole carbon cycle is not well understood. The annual
increase in carbon in the atmosphere is estimated to be only
about 2.3 billion tons. Thus the increased carbon levels being
emitted both from forest clearance and fossil fuel burning are
being absorbed: either by the oceans or by unknown terrestrial
sinks.

Without doubt, a great deal of further research will be required
before the effects of forests on the global climate are properly
understood.

2.8 Forests, rainmakers?

Equally controversial is the effect of forests on local rainfall. There is a widespread belief that deforestation causes a decrease in local precipitation, and that conversely, restoring forest cover will lead to an increase (Goodland and Irwin, 1975; World Water, 1981). If it was indeed true, such an effect would have a great impact on agriculture.

The scientific literature on this subject is far from conclusive. In India, the influence of forests on precipitation has been debated for nearly a hundred years (Singh, 1988). Some studies have reported a decrease in rainfall in certain districts after forest clearance (Warren, 1974), while others have noted an increase following reforestation (Eardley-Wilmot, 1906). A beneficial effect on the number of rainy days per year has also been recorded in some studies (Ranganathan, 1949). However, no clear overall pattern has emerged and it is generally concluded that although there may be some relationship between forest cover and rainfall, the effects on total precipitation are relatively small (Hill, 1916).

A study in the Central Congo Basin found no evidence of any influence of forests on rainfall. It was suggested, however, that forest clearing, by increasing the heat reflectance, might introduce some instability into weather patterns which can be equally important as total rainfall to production systems (Bernard, 1953).

Throughout much of the tropics, most local precipitation is the result of monsoons or major storms generated by large weather systems, or else is caused by moisture-laden air being forced upwards as it passes over hills and mountains. In neither case is tree cover likely to have any dramatic influence on total rainfall.

There are two special cases which merit attention, however, the Amazon Basin and the "cloud" forests.

2.8.1   The Amazon Basin

The Amazon Basin is a horseshoe-shaped plain, open on the east to the moisture-bearing trade winds from the ocean and ringed by mountains and high plateaux on the other sides. Recent studies have shown that in the Amazon Basin the recycling of water vapour by forest vegetation might indeed constitute an important source of atmospheric moisture for rainfall within the basin (Salati and Vose, 1984).

It has been estimated that conversions of 10, 20 and 40 percent of the forest to shrubs and crops would result in annual rainfall reductions of around 2,4, and 6 percent, respectively (Brooks, 1985). These reductions may not seem large, given the fact that

the average rainfall in the region is in excess of 2000 mm.
Nonetheless, since the dry period in the Amazon already has the
forest ecosystem under stress, even a change of this magnitude
might produce irreversible changes in the natural forest (Salati
and Vose, 1984). Even if these changes do not effect the global
climate, the effect on agricultural production in the region
could be disastrous.

2.8.2   Cloud Forests

The second case occurs where persistent moisture laden clouds or
fog are driven through forests or belts of trees by wind. "Cloud"
forests occur at higher elevation on many mountains and are often
made up of unusual plant and animal communities. Worldwide,
cloud forests cover about 500,000 square kilometres, or almost
5 percent of the closed moist tropical forest (Persson, 1974).
"Fog" forests which occur in some coastal areas (sometimes in
areas with normally very low precipitation, such as the coast of
Peru and Chile) act in similar ways. These forest areas can have
a significant impact on a region's hydrological system, and thus
on agricultural production.

These tree barriers strip the moisture from the clouds or fog.
Single trees or narrow belts of trees are most effective. In
closed forests there is a mutual sheltering effect. A study in
Hawaii found that a single _Araucaria heterophylla_ tree added 760
mm of 'horizontal' precipitation in a year to the normal
vertical precipitation of 2600 mm (Ekern, 1964).

This extra moisture is added to the hydrological system and can
add to groundwater and streamflow levels. Because of their
height and the large amount of moisture-stripping surface, trees
are much more effective in this function than other vegetation
types. Maintaining forests in these areas is therefore critical
to local hydrological regimes. Conversely, where wind-driven fog
or persistent cloud phenomena occur in areas that have been
cleared, planting trees can re-establish a water capture system.

2.9 Forests and genetic resources

A final, and important, connection between forests and food
security is their role as reservoirs of genetic diversity.
Though not strictly an environmental link, the fact that forest
environments provide a habitat for a great diversity of plant and
animal species makes them important biological resources.

Forest areas represent the largest single store of genetic
diversity. From the point of view of future agricultural
production, the species they contain - both known and yet to be
discovered - may have a critical role to play in providing the
genetic variation needed to combat the ever-adapting pests and
diseases that prey on food crops. They may also provide a range
of entirely new foods and medicines - of both plant and animal
origin - that could have a major impact on human health and
nutrition.

Conserving these genetic resources for future generations is
being increasingly recognised as both a moral and practical
imperative. The problem is in devising ways of achieving this.

Ex-situ conservation of genetic resources using gene and seed
banks undoubtedly have important roles to play. But there are
limitations to these methods: especially their high costs and
associated technical problems such as genetic drift within
breeding populations. In the foreseeable future, in-situ
approaches in which species are conserved within their natural
habitat will have to shoulder the major burden of genetic
resource conservation.

This means preserving certain areas of forests in an intact, or
near intact state. This poses a set of difficult challenges,
given the many human and economic pressures on forests. Simply
fencing off areas of forest is not possible in many cases, as
huge numbers of people depend on those forests for their
livelihoods. Compromises will have to be made, and ways will
have to be found of combining conservation with sustainable
utilisation of forest resources by local people. For unless
local people have a stake in the survival of forests,
conservation efforts are doomed to fail.

Chapter 3 Forestry and food production

The second chapter explored some of the ways that forests help maintain a stable environment: in the largest sense maintaining the global climate, as well as at the micro-level (e.g. the shade of a tree). The forest environment can, therefore, have an impact on food production: influencing soil, water, temperature and light regimes. Forests and farm trees also contribute directly to food security providing fruit, nuts and other edible foods. These contribute to people's diets in almost all rural areas; for some communities these foods play a major nutritional role. Forests also provide a habitat for a large number of animals, fish and insects which often provide essential additions to rural diets.

Less obvious are the many indirect contributions that trees and forests make to food production. In many livestock production systems, trees are an essential source of fodder, especially during the dry season, and thus contribute to meat and milk production. Mangrove forests provide essential habitats, especially breeding grounds for many fish species, thus helping to maintain fisheries in coastal areas. And as was discussed in the last chapter trees grown on farms can help improve soil conditions.

3.1 Wild foods from the forests

Forests and woodlands, and the wild plants and animals they contained, were once the main source of food for many early hunter-gatherer societies. Over the millennia, with the development of cultivated varieties of wheat, rice and the other staple crops, and the domestication of livestock, man's dependence on forests has declined. Nevertheless, there are a great many rural people who remain dependent on forests for critical portions of their food supplies.

Isolated forest communities exist where wild plants and animals are still the major source of food. In India, for example, some tribal groups depend almost entirely on hunting and gathering in forests and have little contact with the outside world. Similar communities exist in Papua New Guinea and in parts of Africa and Latin America. But while they are the most obvious examples, these are not the only people who rely on forest foods; for many millions of families living outside the forests, forest foods remain an essential supplement to their diet. The issues of who within a community depends most on forest foods, and to what extent are discussed further in Chapter 4.

The array of different foods consumed is vast; it ranges from beetle larvae to nuts and honey. For example, in the arid and semi-arid Sahelian belt of Africa, as many as 800 different edible plant species have been identified (Becker, 1986). One group of agro-pastoralists, the Tswana, use 126 separate plant species and 100 animal species as food sources (Grivetti, 1976).

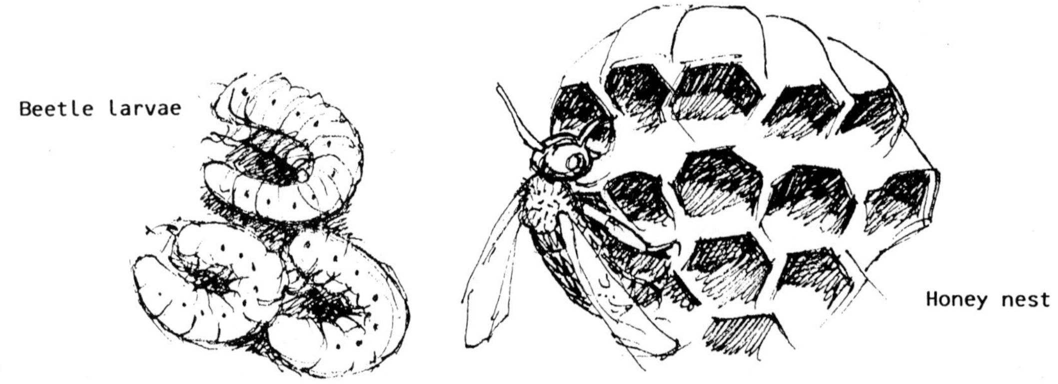

3.1.1   <u>Food from Wild Plants</u>

Several attempts have been made in recent years to catalogue forest food species (FAO 1982; FAO 1983a; 1983b; 1984; 1986a; 1986b). Although a large number of species have been identified with food uses, often this is as far as the information goes. Very little is known about the quantities produced, the seasonality of production, or its variability from year to year. Thus, it is often difficult to assess their relative importance as food sources.

Another factor which complicates the discussion of the relative merits of different forest foods is the pronounced differences in the quality of wild foods depending on varieties, ecotypes

and provenances. The baobab tree, <u>Adansonia digitata</u>, is a good
example; while some trees have soft and tasty leaves that are
highly sought after by local people, with others the leaves are
fibrous and bitter.

Broadly, forest plant foods can be categorized as leaves, seeds
and nuts, fruits, tubers and roots, fungi, gum and sap.
Collectively they add diversity and flavouring as well as
providing protein, energy, vitamins and essential minerals to the
human diet. Some are collected and consumed raw while others
require complex processing before they can be eaten.

<u>Leaves</u>

Wild leaves, either fresh or dried, are one of most widely eaten
forest foods. Typically they are used as a base in the soups,
stews, and relishes which traditionally accompany a carbohydrate
staple. This combination is important because as well as
providing nutrients these wild leafy vegetables add flavour to
otherwise bland foods, and encourage greater food consumption.

The nutritional value of leaves varies widely. Some of the most
nutritious, such as the baobab, contain up to 13% protein.
Others are good sources of Vitamin A, Vitamin C, calcium, niacin,
and iron. Although unusual, the leaves of some species also
contain substantial quantities of fat - for example, <u>Bidens
pilosa</u> (22.5%) and <u>Dracaena reflexa</u> (18%).

Leaves are an important part of traditional diets in many parts
of Africa. In Upper Shaba, Zaire, for example, it was found that
leaves from 50 different tree species were eaten (Malaisse,
1985). Wild leaf vegetables are the most frequently consumed
wild plants in Swaziland, according to another study, with 48

different species being commonly used. More than half the adults
interviewed reported they ate wild leaves at least twice weekly
when they were in season (Ogle and Grivetti, 1985). While another
study found that in Lushoto, Tanzania, wild leaves are eaten at
nearly a third of all meals (Fleuret, 1979).

Boiling fresh leaves in stews is the most common cooking method.
Some leaves, however, are dried and powdered.   In parts of
Senegal, powdered baobab leaves are eaten with couscous. In other
cases leaves may be fermented as a means of preservation. <u>Cassia
obtusifolia</u> leaves, for example, are fermented and used as a
high-protein meat substitute, called 'kawal'.   The fermented
leaves are made into a paste, or are dried and powdered. Kawal
is used in stews and soups which accompany a sorghum porridge
(Dirar, 1984).

Seeds and Nuts

Seeds and nuts generally supply calories, oil, and protein.
Edible oil consumption is low in many developing countries, and
oil is often one of the major household food purchases. Low fat
diets are thought to be detrimental, especially for children who
need energy-dense foods. Fats and oils are also important for the
absorption of Vitamins A, D, and E.

From the point of view of nutrition, the most important nut-
producing species are coconut palm, oil palm and babassu palm.
Coconuts are of central importance in many cultures; on a world
scale, they represent 7% of total fat consumption. Other widely-
eaten species include the sheabutter nut, cashew nut, and
mongongo nut (<u>Ricinodendron rautanenii</u>).

In many parts of the Sahel, the seeds of <u>Parkia biglobosa</u> form
an integral part of the diet. In this region, fermented Parkia
seeds, or 'dawadawa', is an important ingredient of the side
dishes, soups, and stews made to accompany porridges.   The
fermentation process improves the digestibility of the protein
and increases the vitamin content, producing a highly nutritious
food rich in both fat and protein. In parts of Northern Togo,
fermented Parkia seeds are eaten almost every day (Campbell-
Platt, 1980).

Zizyphus spina christi - a wild fruit

Fruits

There are hundreds of species of wild fruits used worldwide. They are mostly eaten raw as a snack food, although some, such as <u>Artocarpus communis</u> (breadfruit), are dietary staples. Many fruits provide a useful source of minerals and vitamins. The fruits of <u>Ziziphus jujube</u> (var. spinosa) are one exceptional example; they contain seventeen times as much Vitamin C per unit weight as oranges.

Rural people are often familiar with a wide range of different fruits. Studies in Swaziland identified 110 edible wild fruit species, of which 13 were eaten frequently by more than a quarter of those interviewed. It was noted, however, that there was considerable variation in fruit abundance and consumption between ecological zones. There were also differences in the amount consumed by different family members; children generally ate the most (Ogle and Grivetti, 1985).

Roots and Tubers

Roots and tubers provide carbohydrates and some minerals. They are used as drought and famine foods not only because they can survive under reduced precipitation, but also because they themselves can be an important source of water. They are also consumed as snacks by children, herders and others who rely on "bushfoods" during the working day. Roots and tubers are also used as ingredients in traditional medicines.

Many roots and tubers require lengthy processing, usually soaking and cooking, in order to be edible. This probably accounts for their use primarily in times of food shortage. In recent years, however, the availability of food aid and commercial supplies may have reduced their importance as a famine food.

Mushrooms

Mushrooms are favourites in many cultures, and are often consumed as meat substitutes. They are good sources of protein and minerals. In one study in Upper Shaba, Zaire, the average protein content of 30 types of edible mushrooms was found to be 22% of dry weight. Here, mushrooms are gathered by women and children, who frequently spend up to two or three hours a day gathering them in the rainy season. The mushrooms are then often marketed (Parent, 1977). Similarly, in the Mae Sa Valley in northern Thailand, many species of mushroom are collected in the rainy season for consumption and sale (Jackson and Boulanger, 1978).

Gums and Sap

Certain types of tree sap can be tapped and made into beverages, which are often high in sugars and minerals. Gums are also used as food supplements and can be good sources of energy. Both saps and gums have many medicinal uses.

In northern Brazil, the Babassu palm is used for making palm wine. The stumps left after harvesting are hollowed out and the sap which collects in the hollow is left to ferment (May et al, 1985a). Similarly, the Palmyra palm (Borassus flabellifera) is widely cultivated in southern India for its sap, or toddy. The sap is tapped from unopened inflorescences, each one yielding up to two litres of sap a day. The sap is either drunk fresh or left to ferment, becoming palm wine.

The Babassu palm –
used for making palm wine

The gum of <u>Sterculia</u> sp. is used as a dietary supplement by the Wolofs of northern Senegal. It is added to soups and stews, and is a good source of Vitamins A and C (Becker, 1983). Similarly, gum arabic produced from <u>Acacia senegal</u> is traditionally an important food for pastoralists, agriculturalists, and hunter-gatherers. Nomads from Mauritania use it to make N'dadzalla, a mixture of fried gum, butter, and sugar. It is also used as a milk substitute when mixed with sugared water, and is often the staple food for gum collectors in the field (Giffard, 1975).

A duiker

3.1.2   <u>Food from Wild Animals</u>

Forest wildlife is the second main category of food derived from the forests. For communities living in the vicinity of forests, natural woodlands and forest fallow areas, wild animals often play a significant part in local diets; in some cases they provide the single largest source of animal protein.

Discussion of food from wildlife has tended to focus on the large game species such as antelope and deer. In fact, in terms of their contribution to the daily diet, these are rarely the most important species. In many areas large game animals have become

rare or inaccessible (as they are protected by hunting bans).
In addition, their meat is often difficult to preserve.

Much more important are the smaller wildlife species. These
include rodents such as the grasscutter, or cane rat (<u>Thryonomys
swinderanus</u>), and the giant rat (<u>Cricetomys gambianus</u>), both of
which are highly popular in parts of West Africa. Squirrels,
porcupines, bats, mice and other small mammals are also eaten,
together with birds and various types of insects, snails, snakes
and other reptiles.

Practices and local preferences vary greatly from place to place.
In some West African communities, for example, children herding
livestock remove ticks from the cattle and roast them for food.
In other cultures, ticks will not be touched. Elsewhere, people
regard frogs as a delicacy, while others would not dream of
eating them.

It is difficult to calculate the extent to which wild meat
contributes to local diets. Hunting of large game animals is
often carried out illegally, and many of the more commonly
consumed foods, such as snails and insects, tend to be eaten as
snacks, with the result that their consumption goes unrecorded.

Some of the most detailed information on bush meat consumption
comes from West Africa, where people's reliance on wild animals
for food is exceptionally high (in part because it is in the
tsetse fly zone). Consumption varies greatly depending on the
conditions of wildlife resources. In areas of Nigeria with no
reserve forests and high population density, one study found that
bushmeat contributed only 7% of the total meat consumed. But in
areas near large forest reserves bushmeat provided up to 84% of
the total meat consumed. Similarly, in Cote d'Ivoire, it is
estimated that 70% of the meat consumed by people in the tropical
moist forest zone is bushmeat; whereas, nationally, it only
provides about 7 % of the total animal protein intake (Ajayi,
1979).

Giant rats - <u>cricetomys gambianus</u>

Springhare is a popular bushmeat in Botswana, some pastoralist communities are estimated to get 80% of their animal protein from wildlife. According to one study, total consumption of springhare is equivalent to the amount of meat obtained from 20,000 head of cattle (Butynski and von Richter, 1974).

In Latin America, wildlife still provides an important source of animal protein in some forested areas. Surveys carried out in the Peruvian Amazon between 1965 and 1973 found that rural inhabitants obtained more than 85% of their animal protein from wild animals and fish (Dourojeanni, 1978). Farmers in northern Brazil's babassu palm region also rely to a considerable extent on hunting for animal protein. The babassu palm fruits are important foods for two large rodents, the pacas and the agouti. Stems of fallen palms are also left _in situ_ in order to attract beetle larvae, which are then gathered and cooked.

As a source of protein and vitamins, most wild animals are comparable to domesticated livestock. Some wild species, however, including various rodents, iguanas and pheasants have a higher protein content. Wild meat also tends to have less fat than domestic meat and can be a good source of iron, Vitamin A and Vitamin B.

Some insects are particularly nutritious. Bee larvae, for example, contain 10 times as much Vitamin D as fish liver oil, and twice as much Vitamin A as egg yolk (Mungkorndin, 1981). Some caterpillars are also very nutritious, and have been likened to vitamin pills (Poulsen, 1982).

Besides providing food, wildlife also represents an important source of income for many families. In sub-Saharan Africa there is a long tradition of trade in bushmeat between rural areas and the major towns, where it is sold as a high-priced delicacy. There are well-established trading links stretching from the hunter at one end, through the processors and transporters, to the retailers who sell the meat to urban consumers. In parts of West Africa, snail collection, preparation and marketing is also big business. Districts blessed with snails look forward to the snail season and the harvest it brings.

Commercial ranching and farming of wildlife for meat and other animal products has been attempted in China, Zimbabwe, Thailand and a number of other countries, in some cases with considerable success. Indigenous game species are often better adapted to local environmental conditions than introduced livestock, particularly in arid areas, and are thus more efficient meat producers. By mixing game species with different grazing habits together, or by combined ranching of game with livestock, it may be possible to make better use of the available vegetation than with a single species. The fact that game farming can be integrated with tourism is another potential advantage.

Some wild animals play an additional role in facilitating the production of food from trees and agricultural crops, through their action as pollinators and as natural predators of insect or rodent pests. By maintaining a proportion of forest cover within farming areas, and providing a habitat for wildlife, the agricultural benefits gained from animals can be preserved at the same time as ensuring a convenient supply of wild food.

Obviously, there may be trade-offs involved. Trees near fields, for example, may be a distinctly mixed blessing to a farmer if they are providing a haven for hungry, seed-eating birds. But in other cases, species such as the grasscutter, which would be a pest if their population was allowed to increase too far, can become a valuable source of food when numbers are kept under control through hunting.

3.2 Food producing trees on the farm

Within settled agriculture, the most widespread direct
contribution of forestry to food production is through food
producing trees on farm and fallow land and around the home. The
extent of this contribution varies widely. It should be noted at
the outset that the boundary between forests and farm lands in
many regions of the tropics is not clear: often forest food trees
are selectively left in farm and fallow areas. At one end of the
spectrum are the sophisticated 'homegardens' found in many parts
of the humid tropics, in which food producing trees provide a
major input to local diets. At the other is the single mango
tree, or other fruit tree, planted outside the house.

3.2.1   <u>Home Gardens</u>

Home gardens are defined as "land-use practices involving
deliberate management of multi-purpose trees and shrubs in
intimate association with annual and perennial agricultural crops
and livestock within the household compounds; the whole crop-
tree-animal unit being intensively managed by family labour"
(Fernandes and Nair, 1986).

Home gardens are found in most ecological regions of the tropics
and sub-tropics, although the majority are concentrated in the
lowland humid tropics. Population densities are generally high
in areas where home garden appear; the average size of a home
garden is usually less than one hectare.

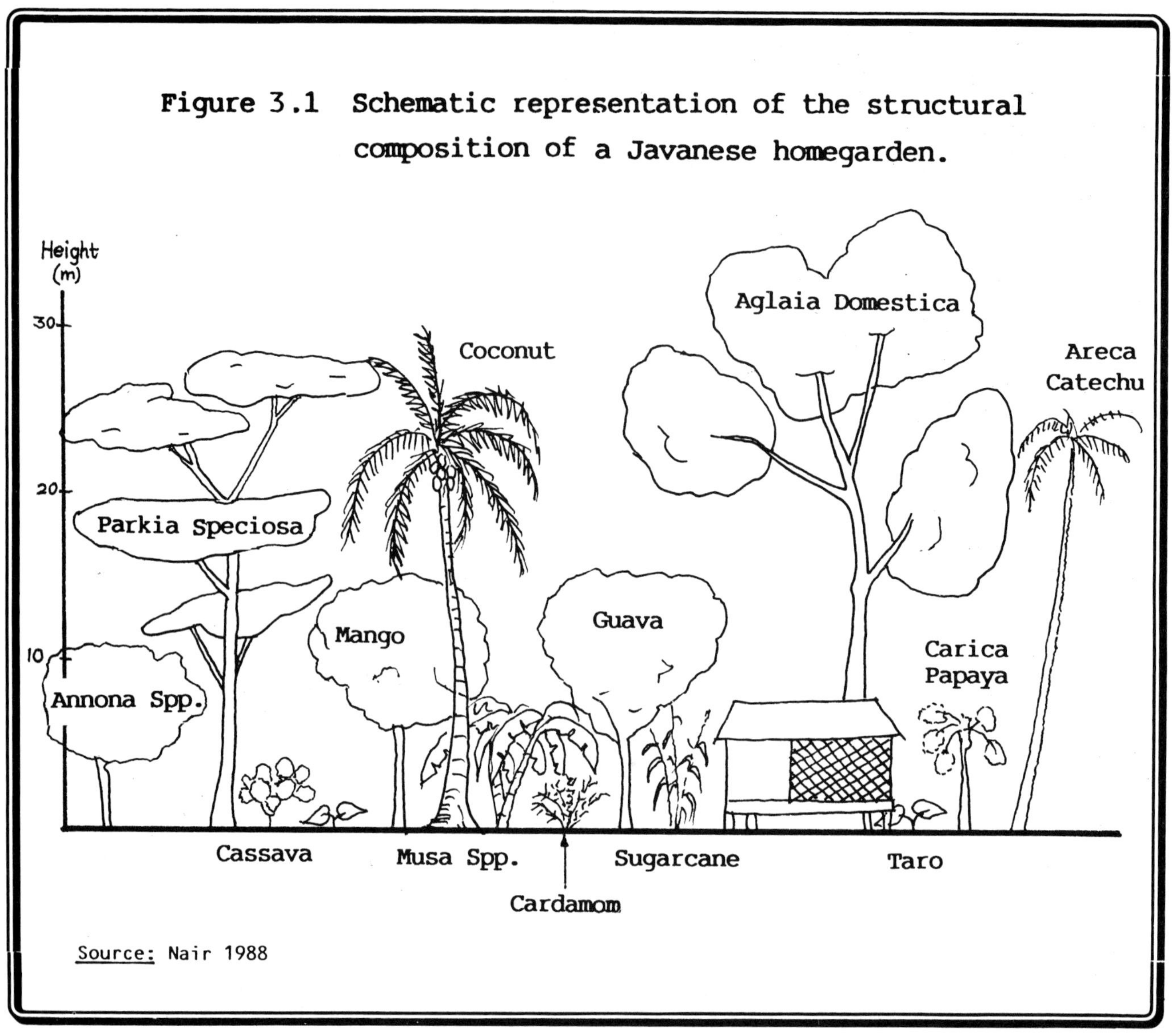

One of the best known cases is the Javanese home garden, which
is schematically presented in Figure 3.1.   These provide an
excellent example of the diversity and complex structure and
function of tropical home gardens.   They have been producing
sustained yields for centuries in an economically efficient,
ecologically sound and biologically sustainable way.

As a general rule, it is fruit trees such as guava, rambutan,
mango, and mangosteen that tend to dominate the Asian home
gardens, along with other food producing trees such as <u>Moringa</u>
sp. and <u>Sesbania grandiflora</u>.   In West African compound farms,
<u>Moringa</u> sp. is common along with other trees that produce leafy
vegetables, as well as trees with fruit for cooking and
condiments that are the most important food producing tree
species.

Food production is the primary function of most home gardens, and
much of what is produced is consumed by the household. When the
tree and other food-producing components are added together, home
gardens can supply a substantial fraction of a family's food
needs.   It is estimated, for example, that Javanese home gardens
provide more than 40 per cent of the total calorific intake of
farming communities in some areas (Terra, 1954; Stoler, 1975).

Another important feature of home gardens is their ability to
produce food throughout the year with relatively low labour
inputs. Crops with different production cycles and rhythms are
combined to provide a year round supply of foods. Although there
are peak and slack seasons for particular products, systems are
designed so that as far as possible there is something to harvest
every day. Any marketable surplus helps provide a source of
income between harvests of other agricultural crops, and a
safeguard against crop failure.

Moringa oleifera with flowers and fruits only 12 months after
sowing the seed in Sudan

3.2.2   Cultivated Food-Producing Trees

Much more common than full-scale homegardens is the practice of
growing a few food-producing trees and shrubs around the home
and on the farm. This happens almost everywhere there is settled
agriculture, although the numbers of trees grown varies from
family to family and from place to place.

Because it currently falls outside the mandate of most forestry
and agriculture departments, tree growing on the farm is very
poorly documented. In terms of nutrition, however, the fruit,
nuts, edible leaves and other foodstuffs produced often supply
an important input to local diets. They also provide a source
of income.

Table 3.1 : Profiles of important food producing species in the tropics

Species	Ecozone/ distribution	Management	Functions/ uses	Common agro-forestry system/ practices involving these species	Other remarks
Areca palm or betal palm <u>Areca catechu L.</u>	Up to 900m, mainly in S. Asia, tropical rain forest zones preferred	Propagation by planting one-year seedlings 2.7m square planting, also in hedges, about 1300 plants/ha, bearing in 5 yrs, up to 60 yrs, responds well to manuring	Seed as a manticatory, edible heart, leaves for thatch in some places, leaf sheath for hats, containers, trunk for wood, seeds also used in veterinary medicine	Cultivated as sole crop or with other crops, usually mixed with cacao and other shade-tolerant perennials, also in home gardens and tree gardens	The crop is not suitable for marginal areas and places with long dry spells
Breadfruit <u>Artocarpus altilis</u> <u>Fosberg</u>	Native to Polynesia, grown all over hot humid tropics, especially in Asia and the Pacific	Propagated vegetatively by root cuttings usually no seed setting, planted 8-10m apart, grown rapidly, bears in 3-5 yrs, needs little care	Mainly grown for edible fruits produced all year round, 700 fruits/ tree/yr, fruits very starchy, vegetable or cooked, biscuits also made, timber useful for farm uses	Usually grown mixed with a large number of other spp. in homesteads, yams usually trailed on trees, offers shade for livestock and crops like taro	Sometimes a staple food in Pacific Islands and the Seychelles
Cashewnut <u>Anacardium occidentale L.</u>	Widely distributed in tropics, Brazil, India, East Africa	Seed propagation, seeds sown at stake, also by vegetative prop. by layering or grafting, about 10m^2 spacing, usually very little aftercare, bearing in 7-10 yrs, up to 50 yrs	Highly priced kernels used in confections and desserts, shell-oil has several industrial uses, cashew apple is juicy and edible, used for wine making, firewood	Cattle grazing under cashew in plantations, tree gardens in small holdings, also in home gardens, used as windbreak and shelterbelt	A very drought-resistant tree, non-synchronized flowering and difficulty in collecting nuts are main problems
Coconut palm <u>Cocos nucifera L.</u>	Coastal areas of the tropics, Philippines India, Sri Lanka, Malaysia, etc.	Propagation by transplanting one year-old seedlings, about 175 palms/ha, square or triangular planting, full bearing from about 8 yrs and continues up to 75 yrs, responds well to manuring	Edible oil from copra (dried endosperm), fruits, drink, leaves for thatch and weaving, trunk for wood, many minor products, acclaimed as "Tree of Heaven"	Many types of crop combinations in smallholdings, intercropping, and multistorey cropping, also grazing under coconuts in the Pacific Islands very common	Most widely cultivated palm alone or with annual or perennial crops, numerous types (dwarf and tall) and cultivated
Date palm <u>Phoenix dactylifera L.</u>	Grown mainly in Arab countries, India, N. Africa, Mexico	Vegetative propagation by basal axillary shoot (suckers), many named cultivars based on fruits quality. Female flowers artificially pollinated	Edible fruit 20-100kg/ tree/yr. Sap for wine, leaves for thatch, weaving, trunk for wood, many minor products, shelterbelts and for sand dune fixation	Grown as overstorey species in oasis and other arid regions, large number of crops grown underneath	It is said to have about 800 different uses
Kola nut (Cola nut) <u>Cola nitida</u> (Vent.) Scott and Endl	Mostly in humid West Africa, also in West Indies, India,Brazil	Propagated by seeds, germination in 7-12 weeks, growth is in flushes, fruiting in 7 years, fruiting up to 80 years, fruits harvested by using knives at tip of long poles	Seeds used as stimulants and beverages, average yield 250kg/ tree but much higher yields reported. Seeds contain 2% caffeine and some essential oils	Interplanted with fruit trees in young ages and with other tree species in adult stages	Fruit is erroneously called "nut"
Mango <u>Mangifera Indica L.</u>	Native to India where very popular but also in S.E. Asia, Africa and tropical America	Propagated by seed or layering and grafting, pruning for shape and induce flowering branches, full bearing in about 8 yrs, bearing continues up to 50 years and more, several cultivars and hybrids	Fruits very delicious dessert, immature fruits in chutneys and pickles, also ripe fruits as preserves, branches for farm construction, timber as firewood, used in dyes	Grown in association with other fruit trees in the backyard, good as border/shelterbelt species, cattle penning in the shade, animal feed or forage	Several forms and types are popular, used extensively on the landscape in India, East Africa

continued:

Mangosteen Garcinia mangostana L.	Southeast Asia. Attempts to introduce to other countries unsuccessful	Seed propagated, seeds have low germination and poor viability. Veg. prop. not successful. Requires shade when young. Bearing in 10-15 yrs, up to 50 yrs. 500-600 fruits/tree/year	A preferred, delicious fruit, eaten fresh. Shell rich in tannins, used for leather tanning and medical purposes	Usually grown mixed with other fruit trees and in home gardens	Tendency to bear only in alternate years, difficulty to propagate, long juvenile phase
Shea butter tree Butyrospermum paradoxum (Gaertn.f) var. parkii	Abundant in Central and West African savannas	Usually propagated by seed, transplanting difficult, about 8m spacing, starts bearing in 12-15 yrs, fruit falls naturally and is then collected	Shea butter extracted from the seed is used as a coooking fat, illuminant, medicinal ointment, shea oil from nuts is used in soaps, candles, cosmetics	Grows in mixed stands with other species in the drier margins of savanna with pronounced dry seasons	Its cultivation is not labour-intensive
Tamarind Tamarindus indica L.	Native to dry parts of Africa, now populat all over Africa, India	Propagated by seed, needs very little care, starts bearing in about 10 yrs, lasts for several decades, fruits are collected from tree or allowed to fall	Fleshy menocarp is eaten fresh or preserved in syrup, seeds eaten as nuts, used as a condiment and flavouring, also produces gums and tannins, firewood, timber good for furniture, foliage and seeds are animal feed	Grows as an overstorey species in many agricultural lands, light canopy and nitrogen fixation are advantages	Grows wildly in drier savannas of Africa and all over India

<u>Source</u> : Asibey 1986

There is a great range of tree and shrub species grown for food.
Some, such as mango and papaya, are popular throughout the
tropics. Others are more localised and are only found in certain
specific geographic regions. For example, rambutan (<u>Nephelium
lappaceum</u>) is common in south-east Asia, while pejibaye (<u>Bactris
gasipaes</u>) is popular in Central and South America (Nair, 1984b).
A summary showing the characteristics, management requirements
and distribution of ten of the most widely grown species is given
in Table 3.1.

Many food-producing tree and shrub species have multiple uses.
Besides providing food, they may be valued for shade or for their
gum and tannin extracts. Leaves may be used as fodder or green
manure, or to provide materials for thatching and handicrafts.
Wood suitable for construction and furniture making can be
obtained from certain trees, and almost all provide a certain
amount of firewood in the form of twigs, prunings and dead
branches.

In some cases these uses are non-competitive. In other cases the
protection and care afforded to many food-producing trees is a
measure of their importance and perceived value to local people.
The cutting of a prize fruit tree for fuelwood or timber is
unusual; where it happens it is often a sign of severe wood
scarcity, or it is because the family needs to raise cash for a
major purchase or unforeseen expense.

3.3 Trees and shrubs as a source of livestock fodder

Another important way in which trees and shrubs contribute to food security is by providing fodder for livestock. In some cases, trees are deliberately planted for their fodder and foliage is cut by hand for stall feeding to animals. More commonly, livestock is allowed to browse on the trees and shrubs growing naturally within grazing areas. Fodder from forest areas, whether browsed or harvested helps sustain livestock production and helps to ensure a year round supply of milk, blood and meat products.

3.3.1   Trees and Shrubs Within Pastoral Systems

Trees and shrubs are particularly important in pastoral production systems. Communities who gain their livelihood from herding animals depend for their survival on an intimate knowledge of their environment. Trees and shrubs are recognised as an essential component within these systems.

There are between 30 and 40 million pastoralists worldwide. Of these, 20-25 million are in Africa, especially in the dry sub-Saharan belt stretching from Mauritania to Ethiopia. The density of woody species within these areas and their relative importance as a source of animal feed is determined primarily by the availability of water. In the driest parts, woody vegetation is scarce and tends to be concentrated along drainage lines and in low lying land where underground water is available. As rainfall increases, woody species become more common.

The range of woody species that are used for animal fodder is extremely wide (Skerman, 1977; Felker and Bandurski, 1979). The fodder they provide, which is collectively termed browse, consists of a combination of leaves, small branches, seed pods and fruits. The importance of browse depends on the type of livestock. Camels and goats are great consumers of leaves and small branches of woody species, while cattle and sheep rely mainly on grasses and annual herbs (Lusigi, 1981).

In many pastoral areas, fodder from trees and shrubs is an indispensable part of livestock diets (le Houerou, 1986; Torres, 1983). This is particularly so during the dry season when the nutritional quality of the herbaceous layer is markedly reduced. At the beginning of the dry season rapid evapotranspiration takes place and the content of digestible protein and B-carotene (necessary for the synthesis of Vitamin A) drops significantly. The dietary energy available also falls due to lignification and the increase in cellulose content at the expense of the more digestible hemicellulose. Animals eating only dry grass would suffer from malnutrition, both because of the shortage of energy and protein, and because of the lack of Vitamin A and essential minerals, especially phosphorus.

Herders recognise the critical role of trees and shrubs. In Sahelian West Africa, most herding groups are well aware of the causes and dangers of vitamin A deficiency, and whenever possible manage their herds so as to avoid pastures with no shrubs or trees.

In part of northern Senegal, it is estimated that during at least 6 months of the year the herbaceous vegetation is not adequate for livestock if used alone because of its high cellulose content and low nutritional quality. Pastoralists are only able to maintain their herds because of the availability of high quality supplements leaves, fruits and seed pods from trees and shrubs (Bille, 1977). As a fraction of total fodder intake during the dry season, browse can contribute as much as 30% of cattle's and 60% of goats' fodder.

A young woman feeds tree leaves to her buffalo in Nepal

3.3.2   Fodder Production and Nutritive Value

A number of attempts have been made to measure fodder production from different components of dryland ecosystems (Trollope, 1981). In general while trees and shrubs are less prolific than annual herbs and grasses, the productivity of grasses tends to be extremely variable, mainly in response to fluctuations in rainfall between seasons, and from year to year. Production of browse from trees and shrubs is much more stable, and because of deeper root systems trees and shrubs are less susceptible to short-term localised fluctuations in rainfall.

The nutritive value of any fodder depends not only on its nutrient content but also on the amount consumed and assimilated by the animal. Although there are a wealth of data on the chemical composition of different types of browse, there is very little information on their effectiveness as animal feeds.

Protein content is one important variable, as this is the main limiting factor affecting liveweight gains of livestock in semi-arid areas (Pratchett _et al_, 1977). Here, browse tends to have an advantage over grass. Comparisons of different types of fodder in the coastal and interior savanna of Ghana, for example, found that browse typically has 2 to 3 times as much protein as grasses, the exact figures varying between seasons.

Though it will generally be the case, fodder with a high protein content is not necessarily a good protein supplement. Measurements of protein digestibility show considerable variations between plant species; with _Prosopis cineraria_, for example, sheep are able to digest only 22% of the protein present, according to one set of measurements, compared to 83% for _Atriplex nummularia_. There are also differences between animals; goats are able to obtain more than twice as much protein from _Ficus bengalensis_ than cattle (Torres, 1983). Thus, simply knowing the species which are commonly browsed and the nutrient composition of their fodder will not give an indication of their value as livestock fodder. The most important feature of forest fodder is that it is available in periods when other feed is unpalatable or non-existent.

A frequently pollarded _Grevillea robusta_ – ideal for fuel and fodder

3.3.3   Improved Use of Tree Fodder

An increasing problem for pastoralists in many areas is the
growing pressure on tree fodder resources. Overbrowsing leads
to reduced regeneration of trees and shrubs, and carried to
extremes will result in their gradual eradication. In parts of
the Sahel this has been an important factor in the decline of
Acacia seyal and A. senegal (le Houerou, 1986). Similarly, in
the central rangelands of Somalia, Yicib (Cordeauxia edulis),
which is the main dry season food of camel and goat herds, is
being overbrowsed and progressively eliminated. This decline is
particularly evident in the 20 kilometres around permanent
watering points (Kuchar, 1986).

There are a number of possibilities for the development and
management of fodder resources both on farms, in rangelands and
forest areas. Trials have been carried out to assess the
potential for increased use of tree fodder as livestock feeds.
Experiments with Leucaena fodder and cattle have shown that for
beef fattening results are comparable with those of concentrated
protein sources (when limited amounts are offered). Milk
production was also improved although the Leucaena tainted the
milk (Jones, 1979).

On rangelands, possibilities also exist for increasing livestock
productivity through greater use of tree and shrub species. A
number of species have been singled out with particular potential
in this context, for example, Opuntia sp. and Atriplex nummalaria
in arid areas of Africa (Kock, 1967), and Prosopis sp. in Latin
America (Felker, 1979).

Improved management of rangelands may also include measures to
control unpalatable woody species such as Calotropis procera, now
common in many degraded Sahelian pastures, especially around
boreholes, and Acacia reficiens, which has made large areas of
Turkana, in Kenya, impenetrable to animals. In both cases,
replacement by more palatable species would add considerably to
the livestock carrying capacity of the area.

In introducing more trees to rangelands, one factor that has to
be borne in mind is the trade-off between fodder production from
woody species and that from the underlying grassy layer. A
balance has to be struck between grasses, which give the greatest
net productivity, and woody species, which are less productive
but are better able to withstand drought. A combination that
gives a high fodder yield during good years, may be disastrous
if production drops dramatically during a dry year.

3.4 Trees and crop production

Shifting cultivators and other farmers who depend on some kind
of forest fallow have long recognized and depended upon the
forest's (and tree's) ability to help improve soil conditions and
thus indirectly crop yields. These effects are most pronounced
in agroforestry systems in which trees or other woody perennials
are grown in close association with agricultural crops. Such
systems, exist in traditional forms in many parts of the world,
as well as in a variety of new, experimental combinations (Nair,
1987a).

Over the last decade, a great deal of attention has been focused
on the development potential of agroforestry systems (Sanchez,
1987). Agroforestry techniques can have a positive impact on
crop production by improving the soil's physical properties,
maintaining soil organic matter, and promoting nutrient cycling
as well as reducing soil erosion and improving the micro-climate
as was discussed in the previous chapter.

3.4.1   <u>Trees and Improvements to the Soil</u>

Under some circumstances, incorporating woody perennials on
farmland can result in a marked improvement in soil fertility.
There are several theories which explain the influence of trees
on soil conditions, notably that the incorporation of trees may
lead to:

* an increase in the organic matter content of the soil
  through addition of leaf litter, decaying roots and other
  plant parts;

* more efficient nutrient cycling within systems, and thus
  better utilisation of nutrients that are either inherently
  present in the soil or externally applied;

* biological nitrogen fixation, and improved solubility of
  relatively unavailable nutrients, such as phosphate, as a
  result of the activity of micro-organisms in the tree root
  zone;

* an increase in the proportion of nutrients that are cycled
  through the plant layer, and therefore a decrease in
  nutrient loss through leaching;

* a moderating effect of additional soil organic matter on
  extremes in soil acidity and alkalinity, and, consequently,
  improved release and availability of nutrients such as
  phosphate and manganese that are sensitive to pH;

* increased activity of favourable micro-organisms in the root
  zone through improvement in the organic matter status and
  temperature of the soil;

* gradual improvement of the physical conditions of the soil
  - in permeability, water-holding capacity, aggregate
  stability, and soil temperature regimes.

The relative significance of these different effects will vary greatly depending on the particular agroforestry system in question, and on local soil and site conditions. Many of these effects also take a considerable time to develop; trees cannot be expected to have a dramatic effect on soil fertility overnight. In addition, while these beneficial effects have been widely assumed, in practice not all of them have been scientifically demonstrated (Table 3.2 presents a summary of the current state of knowledge).

Table 3.2 The Potential Beneficial Effects of Trees on Soils

Nature of Proceses	Processes	Main Effect on Soil	Scientific Evidence
Input processes (augment additions to the soil)	Biomass Production	Addition of carbon and its transfornations	Available
	Nitrogen fixation	N-enrichment	Available
	Rainfall	Effect on rainfall (quantity and distribution) and therefore nutrient addition through rain.	Not adequately
Output process (reduce losses from the soil)	Protection against water and wind erosion	reduce loss of soil as well s nutrients	available
Turn-over processes	Nutrient retrieval/ cycling/release	uptake from deeper layers and "deposition" on surface via litter	not adequately demonstrated
		Witholding nutrient release: this can be regulated by management interventions	available
"Catalytic" processes (indirect influences)	physical processes	improvement of physical properties (water-holding capacity, perseability, drainage, etc.) at the microsite as well as the watershed (macrosite)	available
	root growth and proliferation (enhanced)	addition of (more) root biomass; growth promoting substances; microbial associations	partially demonstrated
	litter quality and cynamics	improvement of litter quality through diversity of plant species; better timing of quantity, and method of application of litter possible	now being increasingly studied in alley cropping and other inter-cropping experiments
	microclimatic processes	creation of more favourable microclimate; shelterbelt and wind-break effects	available
	(bio)chemical/ biological processes (net effects of various processes	moderating effect on extreme conditions of soil acidity alkalinity, etc.	partially demonstrated

Source: Nair 1988

3.4.2   Nitrogen-Fixing Trees

One of most promising groups of tree species from the point of
view of soil fertility are the nitrogen fixers. By virtue of
their ability to capture atmospheric nitrogen and contribute
nitrogen to the soil via leaf litter, or the release of root
debris and nodules (root litter), this group of trees and shrubs
can contribute significantly to the maintenance of soil
fertility.

This ability is already made use of in many traditional
agroforestry systems (Nair, 1987b; Dommergues, 1987). A number
of points need to be borne in mind, however, when considering the
potential for increased use of nitrogen-fixing trees:

* the ability of a particular species to fix nitrogen is
  highly site-specific and depends on climate, soil conditions
  and management practices;

* there is considerable variation in nitrogen-fixing ability
  between different provenances of the same species;

* effective nitrogen fixation requires the presence of the
  appropriate Rhizobium and Frankia strains in the root
  region;

* improvements in nitrogen fixation achieved in the
  laboratory, glasshouse or even in the nursery are not always
  easy to transfer to the field;

* even the least demanding nitrogen-fixing trees require other
  nutrients if they are to flourish, and these needs have to
  be met if their nitrogen-fixing potential is to be fully
  realised;

* the benefits of introducing nitrogen-fixing trees do not
  accrue instantaneously; the effects on soil fertility are
  often cumulative and may take several years to develop.

Thus, the fact that a particular tree is an effective nitrogen
fixer under one set of conditions does not guarantee that it can
be successfully transferred to another. To obtain the full
benefits of nitrogen-fixing trees, a great deal of careful
selection of species and provenances is often required, together
with appropriate management to ensure that the necessary
conditions for nitrogen fixation are provided. Nonetheless,
nitrogen fixers can potentially contribute significantly to
household food security in many farming situations.

3.4.3   Nutrient Cycling in Agroforestry Systems

Agroforestry technologies (whether traditional or newly
developed) improve soil conditions via more efficient cycling of
nutrients. The potential soil benefits of these technologies
depend greatly on local conditions and soil characteristics:
tropical Alfisols and Andepts of moderate fertility appear to be
particularly suited to agroforestry systems (Sanchez, 1987).

In one study in western Nigeria, for example, researchers found that planting Leucaena improved the regeneration of bush fallow on an Alfisol. After three years, during which Leucaena was cut annually and left as mulch, the Leucaena fallow resulted in significantly better soil conditions: higher effective cation exchange capacity and exchangeable calcium and potassium levels, compared to the bush fallow (Juo and Lal, 1977).

The importance of site conditions in determining the effectiveness of agroforestry combinations has been clearly shown, however, in research carried out on alley cropping systems in different parts of the world. This system, which involves growing alternate rows of trees and crops, has proved highly successful in trials conducted by the International Institute for Tropical Agriculture (IITA), in Nigeria, using <u>Leucaena leucocephala</u> interplanted with maize and cowpeas. While alley cropping experiments in Nigeria over six years show a marked improvement in soil fertility (Kang <u>et al</u>, 1985), attempts to duplicate these experiments in a highly weathered, sandy Ultisol in the Amazon basin of Yurimaguas, Peru, were not as successful (TropSoils, 1986).

Although alley cropping works well in moderately fertile soils, current experience suggests that it will be necessary to use inputs such as lime, and possibly phosphorus, to allow successful establishment of alley cropping species and subsequent recycling of nutrients on infertile acid Ultisols and Oxisols (TropSoils, 1986). Further study is required before this particular agroforestry system can be considered widely applicable to the humid and sub-humid tropics. In addition, socio-economic aspects of the applicability of this particular system have not been well studied: i.e. seasonal labour requirements and availability, availability and access to inputs, access to land for tree planting and other tenure issues, and management requirements. In many regions the right socio-economic conditions may not exist for the development of alley-cropping.

3.4.4   <u>Possible Detrimental Effects of Trees</u>

The discussion thus far has highlighted only the benefits that trees provide. The effects of trees on crops are not always positive. If the wrong species are chosen, or if they are planted in an inappropriate manner - for example, at too close a spacing - then trees can have a variety of adverse effects on crops grown in their immediate vicinity. These effects include:

* fast growing trees place a heavy demand on soil moisture; where crop growth is being limited by moisture availability, competition with trees will reduce crop yields;

* uptake of nutrients by trees may deprive adjacent crops of nutrients (although litter fall, and root biomass may compensate for this to some extent in the long run);

* certain trees have adverse chemical and biological effects on nearby plants as a result of acidification, alleopathy, production of toxic exudates, or by providing a habitat for crop pests;

* shading and changes in spectral quality of light can have
  a detrimental effect on the growth of crop species in close
  proximity to trees.

Once again, these effects are highly site-specific and depend not
just on the combination of species being used but also on the way
in which they are arranged, and the management methods used.
Successful agroforestry systems are those that maximise positive
interactions, while minimising the negative ones. In traditional
agroforestry systems the most effective combinations have often
been worked out over generations. With the new techniques
currently being developed, it is only by a great deal of careful
research and on-farm trials that the management and species
optima can be approached. In addition, to the physical problems
associated with incorporating trees into farming, there are many
socio-economic factors which influence the viability of
agroforestry technologies for a particular household or
community. These are discussed in the following chapter.

A three years old Leucaena plantation in the Philippines

3.5 Food production from mangroves

Mangrove forests are unique ecosystems, considered separately here because they contribute to food security in special ways, most notably supporting coastal fisheries. Mangrove forests exist along coastlines in a number of tropical and subtropical areas. Their unique flora is specially adapted to periodic submersion by salt water. They produce a range of plant foods and provide a habitat and breeding ground for a large number of marine animals. In addition they provide a buffer to coastal communities and their cultivated lands against sea storms (as was discussed in the previous chapter).

The total area of mangroves worldwide is estimated at between 160,000 and 170,000 square kilometres (Saenger, 1983). The largest areas are in Brazil, followed by Indonesia, Australia, Nigeria and Malaysia (Hamilton and Snedaker, 1984). They exist in a dynamic state, with their area gradually being extended through the process of siltation, but also being subjected to periodic destruction caused by erosion and violent storms.

3.5.1 <u>Mangroves: Support for Coastal Fisheries</u>

Mangrove forests have a major role in supporting offshore fisheries, and thus in protecting a major food source for many coastal populations. In the Pichavaram mangrove in southern India, for example, 74% of the penaeid prawns caught in adjacent coastal waters use the mangrove as nursery grounds (Krishnamurthy, 1984). In the Gulf of Mexico, it is estimated

that 90% of the commercial catch and 70% of the recreational
catch are dependent on mangrove estuaries for some part of their
life cycle, either during the breeding stage or as larvae,
juveniles or adults. Most of the information on mangrove--
dependent fish and animals focuses on important commercial
species including: bream, mullet, milkfish, mojarras, snooks,
barramundi, seatrout, snapper, drum, croaker, grouper and tarpon
(Hamilton and Snedaker, 1984). There are undoubtedly many little
known species which also depend on these areas and provide a
staple supply of food for nearby communities.

Large quantities of fish, shrimp, oysters, crabs, cockles and
other marine animals are caught in mangroves themselves. The
total annual catch including fish, molluscs, crabs and shrimp is
estimated at around one million tons, slightly over 1% of the
total world fish catch (Kapetsky, 1987).   Besides the
contribution they make to local diets, mangrove fisheries provide
employment for up to half a million people. In most mangrove
areas, the income generated from fishery products is several
times greater than that from forestry.

Oysters, snails, mussels and other molluscs are also cultured in
some mangrove areas. The techniques used range from inexpensive
"spat collectors", where oysters are harvested from aerial
branches, to special rafts as used in the Philippines.

In recent years there has been an increasing shift towards the
use of ponds for rearing fish and shrimp - commonly referred to
as aquaculture. These range from simple ponds which rely on the
movement of the tides to bring in seawater and fresh nutrients,
to more elaborate aquaculture systems involving separate
hatcheries and nurseries, provision of feeds, and use of pumping
to regulate water flows. Even these systems, however, rely to a
certain extent on mangrove areas for nutrients as well as fry
recruitments (Christensen, 1983).

3.5.2   Additional Food Products from Mangroves

A wide range of other food products are also obtained from
mangroves, either cultivated or collected from the wild.

*   Honey is collected from many mangrove forests. Total
    production from wild hives in the Sundarbans in Bangladesh
    forest, for example, was estimated at 263,000 kg in 1983/4
    (Masson, 1984). In Cuba, up to 30,000 hives are moved each
    year, following the flowering season of _Avicennia_, which
    develops in April in the Southwest and lasts until August
    in the North and East of the island.

*   Algae are increasingly cultivated in some countries. In
    Thailand, _Gracilaria_ is grown, the best sites being
    shorelines with sandy bottoms. In the Philippines, algae
    is cultivated as a high value crop for export to Japan
    (Deveau and Castle, 1976).

*   Fruits such as from the Nipa palm are collected from some
    mangrove forests and provide an important contribution to
    local diets.

* Salt is produced by evaporating seawater in many mangrove
  areas. In Pakistan, for example, on the Gulf of Kutch,
  there are 15,000 hectares of saltflats at the back of or in
  depleted low scrub mangroves. Thailand produces over 400,000
  tons of salt a year in saltflats converted from mangroves.
  Some saltflats are turned into fishponds in the wet season
  (Hamilton and Snedaker, 1984).

* Leaves, especially from <u>Rhizophora</u>, are gathered from some
  mangrove forests and provide a high protein animal feed.
  In Iran, the United Arab Emirates and Pakistan, camels are
  traditionally allowed to graze in mangrove areas (Kulkarni
  and Junagad, 1959).

* Emergency foods can be obtained from some mangrove plants.
  <u>Avicennia</u> can be eaten if boiled several times, and Pacific
  islanders use <u>Bruguiera gymnorrhiza hypocotlys</u> baked into
  a sort of bread, after peeling to remove excess tannin. A
  wide variety of traditional medicines are also derived from
  mangrove plants.

3.5.3   Pressures on Mangrove Ecosystems

Although relatively undisturbed mangroves still exist in some
countries, in recent years mangroves have been increasingly
affected by human pressures. Large areas have been converted for
other uses, and many of the areas that remain are being gradually
degraded.

Land reclamation for agriculture is one of the main reasons for
the loss of mangrove areas, although high salinity and the
tendency for soils to become acidified make it a lengthy and
often problematic operation. Urban expansion is another factor;
many major coastal cities are sited partly or entirely on
mangrove grounds - Miami, Panama City, Guayaquil, Sao Luis,
Cotonou, Bombay, Jakarta, and Manila are just some examples.
As these cities have grown, increasing areas of mangroves have
been reclaimed. In addition, conversion of mangroves to
aquaculture ponds has had a major effect in some countries.

Wholesale destruction, or the conversion to other uses, pose an
obvious threat to mangrove areas. What is less apparent is the
gradual degradation of mangroves through various human
influences. In the long run, however, these may be even more
damaging.

These problems relate to a general lack of land use planning for
mangrove areas. Construction of fish and shrimp ponds often
result in excessive destruction of mangrove areas and the
degradation of water and nutrient quality. The uncontrolled
harvesting of firewood and poles, and the lack of replanting pose
another problem. In addition, the various forms of pollution -
urban wastes, insecticides, wastes from sugar cane and other food
processing industries, heavy metals from mining, oil spills,
thermal pollution from power stations - all induce degradation
of mangroves. Finally, the construction of large dams can also
have a significant impact on mangrove survival as they affect
river flows and sediment loads.

In practice, pinpointing the causes of mangrove degradation can
be extremely difficult as all of these factors are
interconnected. While destruction and degradation of mangroves
can be expected to have a variety of deleterious effects, there
are no simple relationships linking, for example, loss of
mangrove areas with reduced offshore fish catches.

There is, however, a clear need for more rational and sustainable
management of the remaining mangrove areas - to preserve the
important role they play both in providing food, and in supplying
other locally important products and benefits.

Chapter 4   The socio-economic aspects of forestry and food security

The previous two chapters have identified some of the forest
"services" and products which contribute to food security:
forests and trees provide critical support to agricultural
production and they provide foods and fodder. Besides providing
food, they also serve as a source of income and capital - part
of which can be used to buy food or invest in future food
production.

This chapter focuses on the socio-economic aspects of forestry's
contribution to <u>household</u> food security. It explores the <u>dynamics</u>
of forestry's contribution to household food security: examining
<u>how</u> households use forest and farm tree resources, and under what
circumstances. It also examines how these uses are changing.

While few studies have focused specifically on food security
issues, it is nonetheless possible to sketch out some of the main
links involved. In terms of household food security, forest and
farm tree resources serve to supplement existing food and income,
fill in seasonal shortfalls of food and income as well as provide
seasonally crucial agricultural inputs and help to reduce risk
and lessen the impacts of drought and other emergencies.

However the picture that emerges is not a uniform one. Trees and
forests play a far greater role in some communities than in
others. For example, forests appear to be especially important
for the rural poor. Neither is the picture static; almost
everywhere, patterns of tree growing and forest exploitation are
evolving in response to changing circumstances, new pressures and
new opportunities.

4.1 The dietary role of forest foods

The last chapter highlighted the great variety of forest and tree foods which are consumed. These food resources are an established parts of the diet for huge numbers of people throughout the Third World. However they rarely supply dietary staples. Nonetheless they often significantly supplement the overall diversity and quality of the diet. In many agricultural communities forest or tree foods are relied upon during the "hunger" season before the new seasons crops ripen. In addition, forests have traditionally provided a source of foods during emergency periods when other foods are unavailable.

4.1.1   Forest Foods as a Dietary Supplement

For some communities, forest foods are a major component in their diets, providing the bulk of their nutritional requirements. This is the exception, however, and is mainly restricted to the few isolated groups of hunter-gatherers still remaining in forest areas. For the vast majority of people, the role of forest foods is a supplementary one; they add variety to diets, improve palatability, and provide crucial vitamins and minerals. Although the quantities involved may not be large in comparison to the main staple foods, they often form an essential component in otherwise monotonous and nutritionally poor diets. Diet diversity is an extremely important element of nutritional well-being, in part because more essential nutrients are consumed, and also because it improves the taste of staple foods thus encouraging greater consumption.

Often, forest foods such as leaves and wild animals are added to soups and sauces which accompany staple foods. For example, the Peuhls from Senegal consume the leaves of _Boscia senegalensis_ year-round in sauces which accompany their grain staple (Becker, 1983). Forest foods are often smoked, dried or fermented making them available over extended time periods;thus helping to ensure a year-round supply of food.

One of the most common uses of forest foods, especially fruit and insects, is as snacks. Most nutritional studies focus their attention on the main meals of the day, and ignore what is eaten between meals. There is therefore very little information on the extent to which snack foods are eaten, or on the nutritional value they impart.

The term "snack" implies their role is somehow peripheral. Yet some studies suggest that snack foods are often consumed in large quantities. Frequently, for example, people eat fruit between meals and while at work, herding, gathering, or tending the fields. A study undertaken in Swaziland found that some types of fruit are regarded particularly as children's food, and are eaten on the way to and from school (Ogle and Grivetti, 1985).

Some forest foods, especially leaf vegetables and wild animals,
are consumed throughout the year by rural households. The most
widespread use of forest foods, however, is in meeting seasonal
food shortages. Many agricultural communities suffer from
seasonal nutrition gaps, or hunger periods. These generally
occur at the end of the dry season and the beginning to middle
of the rainy season, when stored food supplies have dwindled and
new crops are not yet ready for harvesting (Hassan <u>et al</u>, 1985;
Hussain, 1985). Forest and farm tree foods are also valued during
peak periods of agricultural work, when less time is available
for cooking.

In Northern Brazil, the fruiting season of Babassu palm
corresponds to the off-peak agricultural period. The fruits and
kernels make significant contributions to the diet during this
lean period (May <u>et al</u>, 1985b). In Senegal, wild foods are most
commonly used to meet a seasonal food shortage at the beginning
of the wet season. As only two species, <u>Boscia</u> spp. and
<u>Sclerocarya</u> spp., fruit during this hunger period, these are
particularly valued (Becker, 1983).

One study in Zimbabwe has shown how most fruits are consumed
during this annual hunger period. Interestingly, the period of
peak collection and consumption of wild fruits did not correspond
to the main fruiting season. People used fruit to supplement
their diet when it was most needed rather than when it was most
plentiful (Campbell, 1986a).

Seasonal nutrition problems are not just confined to the natural
cycle of dry and wet seasons. Institutional factors can also
cause food shortages. Payment of school fees, for example, is
tied to an administrative calendar. As this may not correspond
with the crop production cycle, it can create a marked seasonal
cash shortage limiting a household's ability to purchase food.
If they are available at the right time of year, forest foods can
also help fill temporary gaps of this kind (Chambers and
Longhurst, 1986).

A papaya tree in Mauritius

4.1.3   The Emergency Role of Forest Foods

Especially in Africa, forests and woodland areas have traditionally played a critical role during emergency periods, such as in times of drought, famine, and war. They provide food when crops fail, as well as yielding products which can be sold to raise cash.

In general, famine foods are different from those consumed during normal years. Many are chosen because they are rich in energy. Their disadvantage, however, is that they often require complicated and lengthy processing. For example, in Zimbabwe, the stems of <u>Encephalartos poggei</u> are soaked in running water for three day, sun-dried, and crushed into a fine powder before being consumed (Malaisse, 1985). In many cases their taste also leaves much to be desired. These characteristics are not entirely surprising. If they were tasty and easy to prepare people would not wait for a famine to eat them; they would be part of the regular diet.

A survey in West Africa found that rhizomes, roots, and tubers are the main sources of energy in times of famine. Various types of bark, pith, buds, sap, stems, leaves, fruit, flowers, and seeds are also eaten. A distinction was observed between periods of crop failures and severe famines: wild forest fruits were found to be useful in the former but less so in the latter. In severe famines, roots and tubers are more appropriate as they tend to be better sources of energy. For example, the leaves and fruit of baobab are commonly consumed during periodic food shortages, while their roots are consumed in famine periods (Irvine, 1952).

In India, Malaysia and Thailand, about 150 wild plant species have been identified as sources of emergency food. The kernels of <u>Aesculus indica</u> and <u>Shorea robusta</u>, and the bark of <u>Acacia arabica</u>, <u>Bombax ceiba</u>, and a number of other species are ground into fine flour to make traditional chapaties (normally made out of wheat or rice flour). The tubers and other underground parts of plants like <u>Arisaema concinnum</u>, and <u>Dioscoria</u> spp. take the place of potatoes and other root crops (FAO, 1983a).

The emergency role of forest food products may be changing with increased commercialisation and food relief programmes. None-the-less, for many poorer people, emergency foods from the forest remain as essential components of their diets in times of hardship. Their contribution to food intake may be small when measured in quantitative terms, but the fact that they can make the difference between surviving an emergency, and succumbing to it, makes the part they play critical.

4.2 Changing diets

The role that forests and trees play in food supply and nutrition
has changed considerably in recent decades, and continues to do
so.   Population growth, privatisation of forest lands and
resources, penetration of commercial markets, conversion of
forest land to agriculture, logging and fuelwood extraction;
these and other forces have put increasing pressure on the
remaining forests. Many forest products that have traditionally
contributed to people's diets are becoming increasingly difficult
to find.

In Botswana, for example, the bushlands have been severely
degraded in many areas. As a result, many traditional wild food
species have disappeared or dwindled dramatically in numbers.
In these regions, the Botswana people rarely use these plants any
more, relying instead on the foods purchased in the commercial
markets. Only at the cattle posts are wild food species still
used to any great extent (Campbell, 1986b).

Reduced diversity of diet is a common trend, and one that
frequently leads to poorer nutrition.   This has been observed
among Pacific Islanders, many of whom have become increasingly
dependent on imported cereals and introduced vegetables, which
have a lower nutrient content than traditional foods.   Fruit and
leaf vegetable use has been greatly reduced, with a consequent
reduction in vitamin and mineral consumption (Parkinson, 1982).

It is often assumed that increased income and assimilation into
a cash economy will raise the nutritional status of rural
populations. In fact, in some cases it is just the opposite. In
some cases the nutritional quality of purchased food does not
compare with that of traditional foods. In other instances
reliance on cash crops makes households dependent on the vagaries
of market prices: a drop in cash crop prices will mean a
household has less with which to purchase foods. Furthermore in
situations where a shift from food to cash crops entails a shift
in control of household income from women to men, household
nutrition may be affected as women are more closely involved with
provision of the household's food (Longhurst, 1985). These issues
have important relevance for forestry projects geared to raising
household cash income and improving household food security, as
the two do not necessarily go hand in hand.

In a study in Bangladesh, for example, the food production and
nutritional status in traditional and modern villages were
compared.   Traditional villages grew two rice crops per year,
while the modern villages grew three. Despite the fact that the
modern village had more food available throughout the year its
inhabitants suffered a higher occurrence of malnutrition. It was
concluded that this was due to a reduced diversity of foods in
the diet, as well as higher energy expenditure (because of having
to produce a third crop), and poorer hygiene.   Although the
modern villagers consumed more rice and wheat, and had a higher

overall intake of calories and protein, the traditional villagers ate more roots and tubers, pulses, vegetables, and fruit. The result was that over the course of the year, the mineral and vitamin content of the diet was significantly greater in the traditional village than the modern one (Hassan et al, 1985).

The role of forest foods in the diet has changed with their diminishing availability and changing tastes and access to new products. In some regions forest foods are rarely consumed and knowledge about their uses is vanishing. This trend is not universal, however.

In some areas forests still provide a readily available source of food and fodder. In addition, commercialisation of rural markets and rapid urban migration have created markets for popular forest foods that previously did not exist. Dawadawa, made from fermented Parkia seeds, is now commonly sold in the markets of Accra, Ghana, for example, well outside its traditional consumption range (Campbell-Platt, 1980). The buoyant market for game meat in many West African towns, and the sale of forest products at the side of major roads, also underline the continuing demand for certain forest foods.

In some countries, people have responded to the declining availability of forest resources by protecting trees or deliberately incorporating them into their farming system. Studies in Zimbabwe, for instance, found that residents in the most severely deforested areas had selectively maintained their favourite wild fruit species (Campbell, 1986a). In other cases, farmers have begun planting fruit trees, both as a source of income, and as a supply of food for the household (Gielen, 1982). Thus, while the availability of foods from the wild may be decreasing, in some cases this is being compensated for by the increased cultivation and deliberate management of desired species.

The impact of declining forest food consumption is not clear. As was noted above, these changes have led to a poorer quality diet, in some cases. Perhaps the worst impact of the loss of forest food resources is that poorer people's food options will be further reduced, especially during seasonal and emergency hardship periods.

4.3 Fuelwood and household nutrition

Fuelwood is the main energy source in most Third World rural
communities. All cooking and most food processing is dependent
on fuelwood. Indirectly, therefore, fuelwood supplies affect the
stability and quality of food supplies. With fuelwood becoming
increasingly scarce in many rural areas, this raises a number of
concerns about the likely impact on nutrition. While there are
few studies that have specifically focused on the links between
nutrition and fuelwood, some of the most important relationships
can been identified.

Fuelwood shortages, for example, may influence the amount of food
cooked. In one extreme case, it was reported that refugees in
Somalia fed their bean rations to their livestock or discarded
them because they could not afford the fuelwood to cook them
(Cecelski, 1984).

Reports from other countries have noted a reduction in the number
of meals cooked per day as a result of fuelwood scarcities. In
parts of Sudan, food is now cooked once a day instead of the
customary three times, according to one study (Hammer, 1982).
This trend may be particularly harmful for children, since, if
the staple food is starchy, a child may not be able to digest
adequate calories in one meal alone.

It is not always clear, however, whether less frequent cooking
results in less food being consumed. Nor is it obvious whether
fuelwood scarcities are the only cause of this decline. Since
wood shortages are often associated with other problems such as
food scarcity, increased workloads, and increased availability
of "fast" foods, a variety of factors may be at work.

A second consideration is that fuelwood scarcity may affect the
quality of food consumed, if it results in reduced cooking time
and greater reliance on uncooked or reheated foods. Eating
under-cooked foods and reheating leftovers can have a serious
impact on disease incidence. This is especially true of meats
because of the danger of parasites, and of tubers and legumes
which need to be cooked properly to destroy toxic components.
One study in Peru found that in one area the consumption of
half-cooked food was common, especially during the rainy season,
and that it had a noticeable effect on the nutritional status of
the families concerned (Alcantara, 1982).

Collecting fuelwood in Ethiopia

It is not only food quality which could be affected by fuelwood scarcity, the quality of drinking water may decline if water boiling is reduced, thus inducing increased incidence of diseases.

Changes in diet may also be associated with fuelwood shortages. Several authors have suggested that an increase in consumption of fast-foods and purchased snack foods may be in response to increasing fuelwood shortages (Cecelski, 1984; Agarwal, 1986). Generally, it is assumed that these foods are of lower nutritional quality than traditional foods, though there is little direct evidence documenting this. It is difficult to distinguish the effects of fuelwood scarcity from other factors associated with changes in dietary habits such as changing cultural values and increased urbanisation and commercialisation.

Fuelwood is also important for food processing often being used to smoke,dry and preserve foods. Food processing is of central importance for food security, as it serves to extend the supply of foods into non-productive periods allowing these resources to be spread more effectively over the year. In the case of commercial food processing (e.g. fish-smoking) if fuelwood is scarce (and thus expensive) this is likely to affect both the availability and price of the final product.

This is one of the problems for the fish processing industry in Kenya and Tanzania. A large percentage of the fish catch from Lake Victoria is currently smoked. Fuelwood scarcities in the region have meant that the cost of processing has gone up, costs which have had to be passed on to local consumers (Mnzava, 1981).

4.4 Forestry and disease

The linkages between forestry medicine and nutrition are
extremely important. Many intestinal diseases, for example cause
malnutrition by preventing the absorption of food by the body.
Disease also debilitates and can affect food production by
reducing labour efficiency during peak periods in the
agricultural calendar.

Forests provide the only medicines available to a large
proportion of the world's population. Many studies have
catalogued the use of medicinal products gathered from the
forests (Heinz and Maguire, 1974). While the effectiveness of
different traditional plant treatments are still under
considerable dispute, a few observations are important. Some
plants contain high concentrations of particular chemicals which
are the base for modern drug equivalents. Secondly, many plants
chosen for their traditional medicinal qualities have high
concentrations of vitamins and minerals which can help counteract
illnesses caused by dietary deficiencies.

As was discussed in the second chapter, forests can influence and
regulate water quality to a certain extent. In addition, fuelwood
supplies the energy for water boiling. Water quality has a direct
influence on disease incidence, and thus people's ability to
absorb foods.

Some trees have properties which can directly affect the quality
of water supplies. _Moringa_ sp., for example, are used by women
in Egypt and Sudan to clarify turbid water. The seeds of the
tree contain natural coagulants which can clear water to tap
water quality in 1 to 2 hours. The elimination of turbidity is
accompanied by a 98-99% elimination of indicator bacteria. Thus
the use of Moringa seeds can provide a low cost water treatment
technology, thereby improving the health of rural communities
(Jahn, 1986).

The fruits of _Balanites aegyptiaca_ and _Swartzia madagascarensis_
contain saponins. These are lethal both to the snails which act
as the intermediary host of bilharzia and to the water flea which
harbours the guinea worm. Planting these species along
irrigation banks, it has been suggested, could do much to prevent
the occurrence of the diseases (Wickens, 1986).

Forests may also have a negative impact on health by providing
habitats for certain endemic disease carriers. Notorious among
them is the tsetse fly, which causes trypanosomiasis in humans
and cattle. Eradication efforts in some countries have involved
the large-scale clearance of natural woodlands, as well as
chemical spraying. The effects have been controversial, however,
since in the process of opening up land for livestock and humans
it may also expose fragile, and previously protected areas, to
rapid environmental degradation.

4.5 Income and employment from forests

Millions of rural people depend on forests for income and employment. For many, the money earned from collecting, selling or processing forest products provides an essential input to family income enabling them to buy food and invest in future food production (e.g. purchase of seeds, or tools).

The particular products involved vary from region to region depending on markets, local traditions, alternative means of employment, and the types of forest resources available in the area. These activities, however, have a number of important characteristics in common:

* they are small in size and are often household-based;

* they are accessible to the poorer sectors of society;

* they are labour intensive;

* they require few capital inputs;

* they provide direct benefits to the local economy.

In the same way that forest foods contribute to family nutrition, forest-based activities usually provide a supplementary source of family income. Similarly, activities follow the seasonal patterns of agricultural cycles and tend to be concentrated during certain periods of the year when labour and the necessary input materials are available. They may also be particularly important in periods of hardship when cash is scarce because of failure of crops or other emergencies.

Two main categories of income-generating activities can be distinguished; those based on gathering forest products, and those centred around the processing of forest products.

4.5.1   Gathering Enterprises

The gathering and sale of forest products is an important economic activity for a great many rural people. A multitude of products are gathered for local, urban, and in some cases, export markets. Since such activity occurs on the fringes of the formal economy, its nature and magnitude is rarely reflected in national statistics. Much of the information that is available comes from anecdotal accounts and localised case studies.

Many studies focus on forest product gathering and trade by forest dwellers (Weinstock 1983, Connelly 1985, IDRC 1980). Many agriculturalists, however, also depend on these activities especially during the off-peak agricultural season. Forest gathering activities are especially important for poorer households in rural areas (Siebert and Belsky 1985).

The collection of rattan has been studied in a number of countries. Derived from a climbing palm (<u>Calamus</u> sp.), rattan provides a source of income for many South Asian people, both forest dwellers and settled agriculturalists (IDRC, 1980). One study in the Philippines found that rattan collection provided an essential income supplement for many farming families, few of whom could survive on their agricultural income alone - especially during drought years (Siebert and Belsky, 1985).

In north-east Brazil, collection, processing and sale of Babassu palm kernels (<u>Orbignya phalerata</u>) is an important source of income for millions of subsistence farmers. The majority of farmers in the area are landless tenants and kernel collection is one of the few ways they can supplement their income. Although most Babassu palm stands are wild, the sale of kernels is generally controlled by the wealthy landlords. Kernel collection and sale corresponds with the slack period in the agricultural calendar, which is the period of greatest cash need. In addition to food purchases, this income often contributes to agricultural inputs (e.g. seeds) for the following season. The palm also provides a multitude of other products including thatch, basketry, charcoal, and food (May <u>et al</u>, 1985b).

The fuelwood trade is an increasingly important source of income for many rural people, especially women. For example, it is estimated that as many as 2 to 3 million people in India are dependent on the fuelwood trade, earning an average Rs. 5.50/day per 20 kg. headload (Agarwal and Deshingkar, 1983). Most fuelwood surveys have focused on the consumption and physical supply of biomass. Only recently have studies begun to address issues such as the income to be earned by rural households in the trade.

Bamboo - a village craft in India

One such study has been carried out in Sierra Leone (Kamara, 1986). Here, it was found that the rural firewood market was concentrated in villages near roads leading to towns. The fuelwood sellers, the majority of whom were women and tended to be the older members of the household, were mostly part-time, selling wood to supplement their household income. The cash earned played an important role in the agricultural cycle. It provided the first cash income from land cleared for rice production. Subsequently, fuelwood selling was concentrated in the off-peak agriculture period, providing cash at a time when food supplies were at their lowest. In one area close to a thriving urban market, fuelwood collection was almost as profitable as upland rice production. This is not typical, however. In most places, in Sierra Leone and elsewhere, the fuelwood trade brings very poor returns.

4.5.2   Processing Enterprises

There is a wide range of forest and tree products which undergo simple processing at the household or small-scale rural enterprise level. A recent survey has been carried out in six countries, looking at the nature and magnitude of forest-based small-scale enterprises and assessing the contribution they make to rural income and employment (FAO, 1987). It was found that the most common enterprises are those producing furniture, agricultural implements, vehicle parts, baskets, mats and other products from cane, reeds and vines. These products are made primarily for the rural market. A variety of handicrafts, however, are also produced to serve urban and sometimes export markets.

Most enterprises are very small; over half the units surveyed were one-person operations, and most depend on family labour. Their average size, and some of the other basic characteristics of forest-based small-scale enterprises, are shown in Table 4.1.

Like forest gathering activities, processing enterprises often operate on a part-time or seasonal basis. They too are dependent on the cyclic demands for agricultural labour, the seasonal availability of forest products as well as the cyclic nature of agricultural incomes-- since the local market for many processed forest products depends on rural people's purchasing power.

As with other small enterprises, those dependent on forests have to be able to respond to market conditions if they are to be successful. There are a number of strategies they can follow. One is to concentrate on market niches in which factory products are not competitive, such as very low cost basic furniture items below the price range of factory products, or high quality hand-carved pieces. Alternatively, they can focus on products in which there is no competitive advantage from large-scale machine production, such as handicrafts. Another approach is to specialise in a particular product or process in order to get the advantages of longer production runs.

Table 4.1   Characteristics of Forest-based Small-scale Industries

Attributes	Jamaica	Honduras	Zambia	Egypt	Sierra Leone	Bangladesh
Proportion of total FBSSIs (%)						
One-person operations	58	59	69	69	–	36
Production at home, not workshop	52	72	81	76	–	–
Rural location:						
– Enterprises	88	100	96	80	99	97
Employment	79	100	95	65	96	–
Women's share:						
– Ownership	32	10	12	65	–	(3)
– Labour force	30	6	12	31	–	21
% family members in						
– Labour force (No)	82	51	86	89	(41)	73
– Hours worked	68	57	–	89	34	–
Mean Values:						
No of workers per enterprise	2.2	2.2	1.7^1	1.9	1.8	3.8
Total investment (US$)	3030	10555	–	–	431	255
Hours worked annually per worker	990	1247	1205	1712	2004	836
Annual production value per firm (US$)	4979	2536	–	1501	1384	2362

[1] The number of hours per worker for Zambia is estimated from the one-visit survey.

<u>Source</u>: Fisseha 1987

The small-scale furniture industry in Egypt provides an interesting example of specialisation. Even the manufacture of items such as chairs is distributed between different units specialising either in particular parts such as legs or seats, or in different stages in the production process, such as primary processing, assembling or finishing (Mead, 1982).

In northern Thailand, small village-based entrepreneurs have taken advantage of the improved roads in their region to truck the furniture they produce to towns or busy roadsides where they assemble and finish it for sale. In this way they compete effectively with furniture from large urban producers and have expanded their markets (Boomgard, 1983).

4.5.3   <u>Employment in Forest-Based Activities</u>

One of the greatest contributions of forest based enterprises to local economies is the employment afforded a great many rural people. Although the absolute numbers of people involved are not high in relation to the entire rural population, they form a large share of those employed outside agriculture. These activities often provide seasonal employment in periods when few other options are available, especially for the rural poor.

Several studies have attempted to estimate the economic importance of forest-based gathering and processing enterprises in India. In many rural areas, income generated from these sources is a key component in the rural economy. In the north-east State of Manipur, for example, an estimated 87% of the population depends on income generated from forest products. Some 234,000 women in this region are engaged in forest product collection activities.

Elsewhere in India, collection of tendu leaves (<u>Diospyros melanoxylon</u>) is an important dry season employer, especially for tribal groups. As many as 7.5 million people are estimated to be involved. In the areas around the forest, these leaves are used for wrapping 'bidi' cigarettes, which itself is a major cottage industry worth more than $100 million a year and employing a further 3 million people. Over the whole of India, more than 30 million people are thought to be involved in forest-based income earning activities of one kind or another (Cecelski, 1984). In south east Asia, at least half a million people are employed in collection, processing and small-scale manufacturing of rattan products. The trade of unprocessed rattan alone is estimated at $50 million annually (IDRC, 1980).

Women figure prominently as owners as well as employees in forest-based enterprises in some countries. In Jamaica, for example, 32% of the enterprises are owned by women, and women make up 30% of the labour force. There appear to be clear distinctions, however, between the types of enterprises involving women and men. In Zambia, women are owners of a large share of the enterprises involved in broom making, bamboo processing and twine and rope making; but they are rarely involved in carpentry or furniture making (FAO, 1987).

The fuelwood trade is often dominated by women. In Sierra Leone, 80 percent of the urban firewood sellers are women (Kamara, 1986). In a survey of women fuelwood collectors in Gujarat, India, it was found that 70% of women collected fuelwood for sale for more than 25 days of the year (few collected wood during the monsoon season). Most of the income derived was used for buying food (Buch and Bhatt, 1980).

In some regions men are becoming more involved with the fuelwood trade as the distances to be covered increase, and because few rural women have access to the donkeys, trucks or other forms of transport needed to carry wood over long distances. This shift in roles of fuelwood collection may free women of one of their most tiresome chores. But at the same time it may deprive them of an important source of income.

Women play an important role in the collection and processing of Babassu palm fruit in Brazil. While both men and women gather the fruit from the wild, it is the women who process the fruit and kernel oil (May <u>et al</u>, 1985a). Similarly, in Sierra Leone, women are responsible for processing oil palm kernels which are gathered from the wild by both men and women. Much of the income generated from the sale of palm oil goes to men; women, however, retain some of the kernels to earn money for themselves.

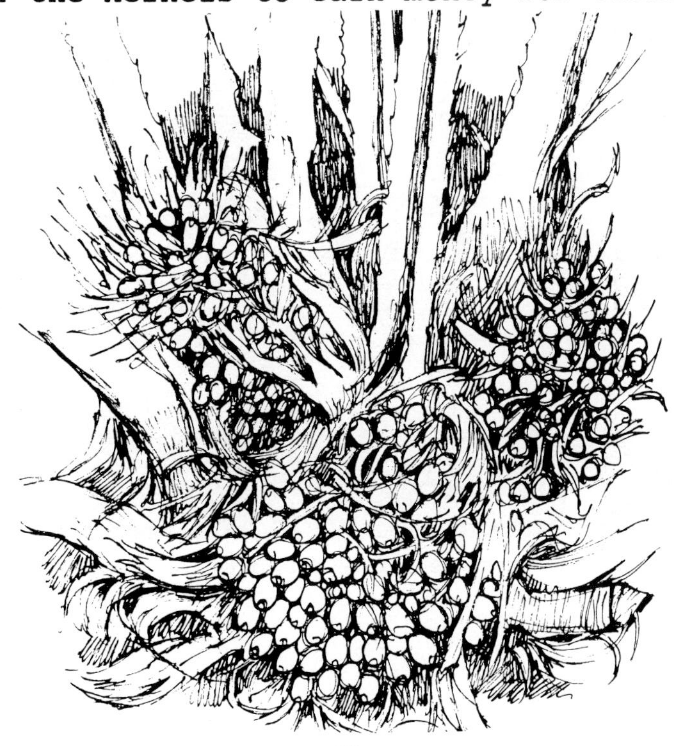

The fruits on a
young oil palm

Because women generally have less access to land and other income earning activities than men, income derived from the sale of forest products is often particularly important to them. The fact that gathering of forest products can often be combined with fuelwood collection, fetching water and other routine activities is one advantage. It also helps that processing can usually be performed at home, allowing women to combine these income earning activities with other household chores.

From the point of view of family nutrition, women's income is often particularly important. Some studies have compared women's and men's spending patterns and have found that women tend to spend more money on basic food supply. Nutritional status is therefore more directly dependent on women's income than men's.

Time constraints are often one of the main factors limiting women's involvement in income generation from forest-based activities. Fuelwood shortages are also a problem for many women involved in fish-smoking, beer-making and other food processing activities that rely on fuelwood. From a food security standpoint, one of the most harmful effects of fuelwood scarcities may be the added burden placed on women's time and therefore limitations to their income earning potential (Ardayfio, 1985).

Fish smoking using traditional kilns - Cote d'Ivoire

      <u>Contribution to Household Food Security: the Role of Forest-Based Income</u>

Income earned from forest-based activities contributes to food security in a number of ways. Most obvious, is the availability of cash for food purchases, especially during hardship periods. In addition, these monies are sometimes invested in agricultural assets such as livestock, tools or land. In this sense, forest resources offer poorer households a means for investment in their future; providing an opportunity to escape from the cycle of poverty.

One of the advantages of small-scale forest-based enterprises is that the benefits accrue directly to the household concerned. For many families, a significant percentage of their income is generated through forest based activities. In north-east Brazil, for example, an average of 25% of household income (including non-cash income) comes from babassu palm kernel gathering and processing during the dry season (May <u>et al</u>, 1985b).

In some areas, collection and processing of forest products has taken over as the main income-generating activity. In one study in Sierra Leone, 18.6% of farmers interviewed said they considered non-agricultural enterprises - which ·included processing activities, fuelwood collection, hunting, fishing, palm wine tapping, and handicrafts - to be more important than farming (Engel <u>et al</u>, 1985).

Palms yield multiple types of products

Hunting for trade in game meat is a particularly lucrative activity in some countries. In Peru, a skilled hare hunter can reportedly earn $1350 a month compared with the agricultural labourer's wage of $100 a month. In Ghana, a single grasscutter sells for more than twice the daily minimum wage in rural areas, and as much as seven to thirty times this wage in Accra. A successful farmer-hunter can therefore earn more from hunting than from agricultural production (Asibey, 1987). Trends in bushmeat prices in Ghana compared to beef and mutton are shown in Figure 4.1.

Table 4.2 Urban Consumer Prices for Meat in Ghana

	Beef		Mutton		Bushmeat	
	Kumasi	Accra	Kumasi	Accra	Kumasi	Accra
1980	22.09	40.88	23.09	NA	78.15	83.95
1981	52.51	47.84	52.83	NA	81.90	144.00
1982	85.51	83.64	88.57	87.56	48.56	180.48
1983	165.00	135.75	150.91	150.33	125.73	373.48
1984	234.17	239.00	234.17	252.67	223.71	453.08
1985	283.94	276.53	305.00	453.15	299.98	510.61
1986	270.41	271.87	260.04	255.96	349.45	684.64

Source: Asibey 1987

The income earned from gathering and processing forest products is of particular importance to the rural poor. In many societies, local people have traditionally had "open" access to forest resources. Poorer groups within the community have thus been able to exploit the forests for food, fuel and other marketable products, and have tended to rely on these for a greater proportion of their income and basic needs than those in higher income groups. Similarly, because of their low investment requirements, small-scale forest-based enterprises are often more accessible to the poor than other income-earning activities.

In the Philippines, dependence on rattan collection has been shown to be linked to income. While poorer families rely on rattan collection and other forest-based employment for their regular income, better-off farmers use rattan mainly as an emergency source of income in times of poor harvest, or other emergencies (Siebert and Belsky, 1985). The same trend was observed in comparisons in Korea, Taiwan, Thailand, Sierra Leone and Nigeria. In all cases, the poorest households with the least land relied most heavily on off-farm income earning activities (Kilby and Liedholm, 1986).

While forest based activities provide numerous opportunities for
the rural poor, some studies suggest the earnings vary
substantially from one activity to another. A study in Tanzania
revealed that returns to labour varied from well below the
minimum rural wage rate for mat-making to several times this
standard wage for carpentry (Havnevick, 1980). In this case
access to markets was a crucial factor determining the
profitability of different activities.

The implications for household food security are unclear: as
women are predominantly involved in craft activities these
findings suggest that household nutrition may suffer as women's
incomes are directly correlated with nutritional well-being. On
the other hand, the benefits derived from products produced to
meet domestic needs may allow a household's cash income to be
spent on other products such as food.

The returns to labour for many forest-based activities are
marginal. In addition, markets for products may be vulnerable to
introduced substitutes. Thus, although forest activities provide
some source of income for a great many rural poor, activities
which are dominated by the poor and women often earn the lowest
returns. These activities may not be sustainable in the sense
that they will be abandoned if other opportunities arise or if
substitute products cause a market collapse.

Certainly, there is not yet enough information to accurately
measure the impacts of marginal returns for forest-based
activities on food security. It is clear, however, that some are
likely to provide more secure and renumerative sources of income
than others.

4.5.6   <u>Constraints to Further Development of Forest-Based
         Enterprises</u>

Small-scale processing and gathering enterprises based on forest
products face a range of problems. Being small, they tend to be
more susceptible to fluctuations in market conditions and
shortages in raw materials. The range of problems that are
encountered by these enterprises can be summarized as follows:

   * insecure markets due to low rural incomes, seasonality of
     production, poor market information, lack of access to urban
     markets, and external competition;

   * raw material shortages, often compounded by wasteful
     processing, restrictive regulations, poor distribution, and
     lack of working capital;

   * lack of access to appropriate technology in the form of
     suitable tools and equipment with which to improve
     productivity;

   * shortage of finance, in particular working capital;

   * managerial weaknesses, which serve to worsen all the other
     problems;

* lack of organisation of the enterprises in a manner which
  enables them to make effective use of available support
  services.

Market forces play a major part in determining the success of
small enterprises. Their position can be eroded by competition
both within the small enterprise sector, and with their larger
counterparts. Due to very low capital and skill requirements for
entry into many small-scale processing activities, it is all too
common for many more production units to exist than can be
supported by the local market. The resulting competition leads
to high failure rates and prevents profitable operations emerging
which can generate enough surplus to be ploughed back into
improving and expanding the enterprise.

The instability of rural markets is another threat to small
enterprises. Incomes, being agriculture-based, have a short peak
during which demand may exceed their capacity to supply. The
resulting supply gap provides an opportunity for larger
suppliers. Lack of working capital is a major barrier preventing
small enterprises from stocking adequate productive inputs to
smooth out seasonal fluctuations in their markets.

Improvements in rural infrastructure which enable products from
outside to be sold in rural markets, and changes in rural market
demands with rising rural incomes, also put small enterprises
under increasing competitive pressure. Thus, factory-made
furniture tends increasingly to displace traditional furniture
made by local artisans. Similarly, bags and mats made from
synthetic materials take over from similar products made by hand
from natural raw materials.

Shortages of raw materials pose a major threat for processing as
well as gathering enterprises. Often this is related to
unselective felling by contractors which fails to preserve unique
species or varieties. Sometimes, the problem is that a specific
type or quality of wood, cane, or other raw material is depleted.
This may be because it is being selectively extracted by large
scale industries, or it may be a result of uncontrolled
harvesting by small enterprises. Almost always, it is the poor
who are most seriously affected as they are the ones who depend
most on income from forest products, and who have least
bargaining power.

In some regions, commercialisation of forest products has
resulted in over-exploitation of forest resources as markets have
expanded. The increased profitability of rattan collection, for
example, has led to its depletion in many regions; where once
it was easily collected it now takes longer trips to gather less
material. Similarly, in parts of West Africa wildlife resources
have been severely depleted because of the increased demand for
gamemeat from urban markets.

4.6 Farm trees: contributions to household food security

Farm trees provide many of the food security benefits associated with forests: providing foods, fuel for cooking and food processing, fodder and marketable products, as well as some of the "environmental" services to food crop production discussed in Chapter 2. At the same time tree cultivation draws on the resources of the farm household and imposes costs of various kinds.

In the previous chapter ("Forestry and food production") the discussion focused on the physical linkages between trees and food crop production describing how trees are or could be integrated in farm systems to increase food production. This section explores the socio-economic conditions in which tree growing can benefit household food security highlighting the linkages between trees and the farm economy, the factors involved in farmer decisions for or against tree growing, and the impact of tree cash crops on household food security.

There are many factors which determine the need and possibilities for tree growing. Farmers have historically protected, planted and managed trees on their lands in order to maintain supplies of products no longer available from natural forests. In addition, trees may be retained to maintain soil productivity, or are grown on sites unsuitable for food crops.

The advantages or disadvantages of tree growing are also determined by economic factors such as the availability of land, labour and capital; subsistence needs and market opportunities. Tree growing is also influenced by cultural factors: e.g. land tenure, attitudes towards communal forest management, and status symbols.

The following discussion examines farming systems where trees are major components in order to identify both the contributions trees make to household food security and the economic considerations that encourage farmers to adopt them.

4.6.1   Home Gardens: Intensive Tree Management

Among traditional tree growing practices, home gardens are one of the systems that have been studied in most detail (see section 3.2.1). In Java, home gardens are prominent features of traditional farming systems, especially in regions of high population density and decreasing availability of crop lands. With growing population pressure, the proportion of land under home gardens has been increasing, in some cases to up to 75 percent of the cultivated land area (Stoler, 1978). Access to rice land has meanwhile declined, and a large proportion of farmers now have no rice land, or not enough to produce their basic requirements.

As a result, home gardens are cultivated more intensively, with
more annual crops being introduced to provide additional food and
income. Labour inputs increase; the labour inputs in small
gardens are reported to be on average three times higher than in
larger gardens (Soemarwoto and Soemarwoto, 1984).

Another way of intensifying the use of home gardens is to
increase the value added from home garden produce. For example,
some of the poorest farmers have shifted from producing just
fruit from their coconut trees to producing coconut sugar, a
highly labour intensive process which, though yielding only low
returns to labour, increases returns to coconut bearing land
(Penny and Singarimbun, 1973).

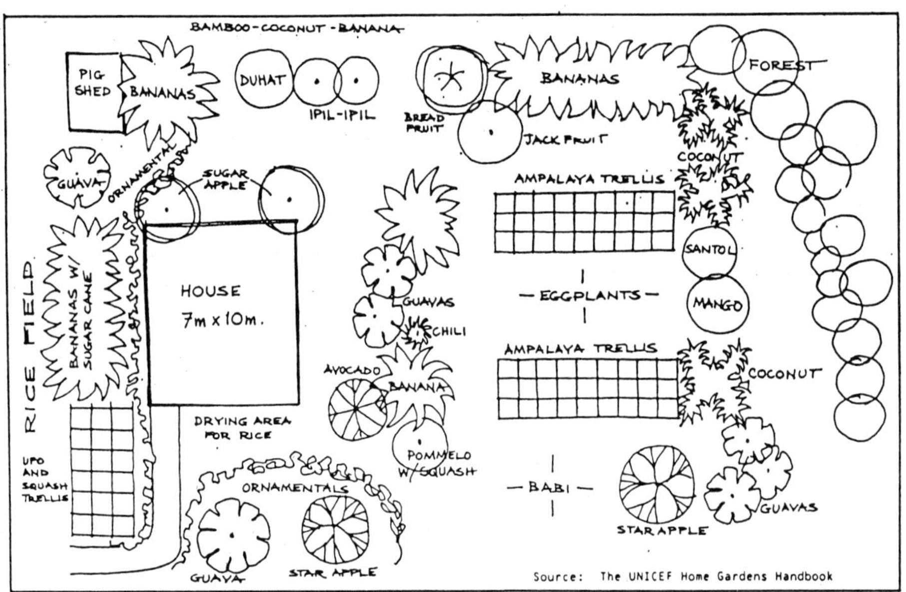

As land holding size continues to decline, income is increasingly
sought from off-farm employment. At this stage, trees and other
perennials requiring only low labour inputs come to form the main
components of the gardens allowing farmers to cultivate their
land while seeking off farm work (Stoler, 1978).

Similar trends have been observed elsewhere. In southeast
Nigeria, for example, farms typically comprise a mixture of
fallow, outer and inner fields and permanently cultivated
compounds around the household. These compounds contain a
variety of tree species, including oil palm, raffia palms,
coconut, banana and plantains intercropped with cassava, yams
and other arable crops.

As pressures on the land heighten, the proportion of land under
compound systems increases, as does the density of both tree and
arable crop cultivation within the compound areas. Compared to
the fields, yields in monetary terms from compounds are five to
ten times as much per hectare, and returns to labour four to
eight times higher. With increasing population density, compound
areas account for up to 59 percent of crop output and a growing
proportion of total farm income. Livestock also become an
increasingly important part of the compound system, providing
food, income and manure. As population density continues to
increase, however, yields and returns to labour eventually
decline to the point at which farmers have to turn to non-farm
sources of income (Lagemann, 1977).

The overall picture, as in Java, is one of farmers responding to decreasing land availability by moving to greater dependence on agroforestry systems. Initially this is because these permit more efficient land use and higher returns from labour than alternative land uses. When pressures on the land increase further, to the point where income has to be generated mainly from off-farm employment, the strategy changes. Agroforestry systems are retained, but in a modified form which allows management and labour inputs to be reduced.

4.6.2   <u>Trees as Cash Crops: the Case of Farm Woodlots</u>

In several countries, farmers have taken up growing trees as a cash crop on land that was previously used for agriculture. Their motivation has been the prospect of earning greater income compared with other uses of the land. Growing trees as cash crops is particularly important for poor farmers. In many cases their resources are too limited for them to meet their basic food needs through crop production, and they are forced to income earning off the farm. In these situations where farmers have little time for crop production low input tree crops may provide the best way of keeping land in use. In addition, trees provide a measure of insurance: they can be harvested in times of emergency cash needs. For poorer farmers, the reduction of risk may be an important consideration.

There has been a recent expansion of tree growing as a cash crop in parts of Kenya (World Bank, 1986). The main species grown are eucalyptus, which is used for poles, and black wattle, which is sold for poles, charcoal, fuelwood and sticks for "mud-and-wattle" construction. Markets for these products - as well as for pulpwood and saw timber in some places - are growing strongly, with farm level production accounting for a large part of the supply.

In these regions, tree growing tends to be practised by poor farmers unable to meet their basic food needs from on-farm production. For some it has become their a principal source of farm income. In parts of Kakamega District, where average farm size is only 0.6 hectares, as much as 25 percent of the land has been planted with eucalypt woodlots (van Gelder and Kerkhof, 1984).

What is surprising at first sight is that gross income per hectare from tree growing is considerably lower than from agricultural crops. Other factors, however, are involved. Alternative crops often require substantial investments, at levels which many farmers cannot afford; trees, by contrast, require very little. Tree cultivation also requires less labour. Here, this is particularly important as widespread outmigration of men seeking off-farm employment has caused a shortage of household labour. In areas where markets for tree products are good, returns to labour from pole production have been estimated to be some 50 percent greater than from maize production (World Bank, 1986). This example illustrates that tree growing is a rational use of resources for poor farmers who need to devote a substantial amount of their time to non farm employment.

Trees are also successfully grown as cash crops by hill farmers
in Haiti. In this case there was already a well established
market for woodfuel and poles, and a strong cash crop tradition.
Most farmers also owned their land. It was hoped that
incorporating trees on their farms would also help them control
the serious erosion problems they were experiencing.

Since 1982 approximately 110,000 farmers have planted more than
25 million seedlings. Patterns of planting vary considerably from
farmer to farmer, but increasingly they have moved from species
suited only to fuelwood and poles production to multi-purpose
species, and to intercropping trees with agricultural crops such
as maize, sorghum and beans.

Surveys of participating farmers indicate that they perceive the
increased income earning potential as being the main benefit of
mixed tree/crop systems. They are also influenced by other
motives. Many plan to use their trees as a form of savings and
value the fact that they can draw upon such savings by harvesting
the trees at a time of their choice. In an area subject to
drought, trees are seen as being less susceptible than crops to
failure, thus reducing uncertainty. With 81% of those
interviewed having to employ agricultural labour, and constrained
by lack of cash in doing so, tree cropping is attractive as a
lower cost use of land. Growing of trees may thus enable poor
farmers to increase the amount of land they are able to work
(Conway, 1987).

Perhaps the best known case of cash crop tree growing is in
India, where large numbers of farmers have adopted tree growing
as an alternative to agricultural crops. Studies of the motives
behind farmer decisions have been made in a number of States
(Skutsch, 1987; Arnold et al, 1988; Tushaar Shah, 1987). In
all cases tree growing is occurring where there are strong and
expanding markets for poles, pulpwood or other wood products.
The main reasons farmers give for switching to trees are as
follows:

   * the lower labour inputs needed with trees, which reduces the
     cost of hired labour and the problems of labour management;

   * the minimal annual operating costs once the trees are
     established;

* the lower water requirements once trees are established and
  their greater resistance to drought, which reduces the risk
  of crop failure;

* the fact that trees provide a way of accumulating a low-risk
  capital asset.

For these Indian farmers, many of whom - though not all - are
large farmers, growing trees as a cash crop offers a number of
advantages. From their point of view, switching to tree growing
may increase their income and therefore indirectly improves their
food security position. The effect on landless families in the
same area, however, may not be so advantageous. Concern has been
expressed that cash crop tree growing is harming the poorer
members of the community by reducing the work available for
agricultural labourers. In practice, reliable data on employment
effects is hard to find. Extra employment from wood processing
activities may make up for at least some of the jobs lost on the
farm. But if the net effect is that substantial job losses are
occurring, then the income and food security position of
wealthier farmers may be improving at the expense of poorer
groups.

4.6.3   Management of Forest Fallows

The previous two examples of farm tree management, home gardens
and fuelwood lots, help to illustrate some of the economic
factors which influence a farmer's management options. These are
systems in which trees are intensively managed often in response
to high pressures on farmer's land or labour resources.

Shifting cultivation, and other farming systems dependent on a
forest fallow, also evolve in response to increased pressure on
resources. In its traditional form, shifting cultivation (swidden
agriculture) is a highly efficient use of farmer resources.
Family labour is the main resource available to the shifting
cultivator. Where there is sufficient land to support fallow,
no other farming practice will produce a higher return to labour
without inputs of capital. The fallow vegetation maintains soil
productivity, and the process of clearing and burning provides
conditions for crop cultivation requiring minimal inputs for soil
preparation and weeding. Though cultivation periods could be
extended by increased weeding, it is easier to clear and burn a
new area. Similarly, yields per hectare could be increased by
more intensive cultivation, but at the expense of lower output
per unit of labour. As long as they can satisfy their production
objectives through less labour-intensive methods, farmers will
logically stick to them (Rambo, 1984; Raintree and Warner, 1986).

As access to land declines, so too does the sustainability of
traditional methods, farmers eventually start to intensify
agricultural practices (Olofson, 1983; Raintree and Warner,
1986). These are usually small incremental changes involving
increased inputs of labour, and sometimes of capital - in the
form of fertiliser or herbicides. In some instances the
evolution away from shifting cultivation may move away from tree
cultivation altogether, but it can include tree management.

A widespread practice at an early stage in this process is to enrich the fallow by encouraging or planting tree species which either accelerate the regeneration of soil fertility or produce outputs of subsistence or commercial value. The cultivation of _Acacia senegal_ as a fallow crop in Sudan is an example of a species that does both; it is leguminous and it produces gum arabic for sale, and fuelwood, fibre and other products for use in the household. Other examples include the management of the Babassu palm for both commercial and subsistence products in conjunction with shifting cultivation over large areas of Northeastern Brazil (May _et al_, 1985a), and the planting of rattan as a commercial crop in the swidden cycle in Borneo (Weinstock, 1983).

Acacia senegal

As pressures on land force the transition towards continuous cultivation, various forms of intercropping may be adopted. By incorporating soil enriching species with food crops, these practices reproduce the functions of fallow. Numerous examples of such continuous fallow strategies are to be found, such as the maintenance of _Acacia albida_ in cultivated areas of the Sahel.

The intercropping of _Sesbania sesban_ with maize in parts of Western Kenya is another interesting example. When the maize is shaded out after about three years the Sesbania is left as a fallow crop for one to two years, and then cleared and used for fuelwood. The cycle is then repeated. Over a ten year cycle it is estimated that maize production per hectare is less than half the monocropped maize yield. The advantage is, however, that it requires less than half as much labour and gives higher maize

yields per unit of labour input - in addition to the fuelwood and
soil protection benefits it provides (World Bank, 1986). In this
situation labour is the constraining factor for food production-
thus once again farmers are responding to interrelationships
between resource availability and production goals.

These examples from three quite different farming systems serve
to illustrate the complex nature of the decision-making process
for a farmer. The availability of resources - in particular,
land, labour and capital - have a crucial bearing on what
management strategy will be the most effective and what role
trees can most usefully play. Market opportunities for farm
products, and the availability of off-farm employment also have
an important influence.

4.6.4   Farmers' Incentives for Tree Growing

It is clear that trees are grown by farmers for many different
reasons. Farm trees can contribute significantly to household
food security: providing foods, agricultural inputs, soil
fertility and a source of cash income. An understanding of how
and when trees can best be exploited by farmers is essential for
forestry programmes geared to improving household food security.

Comparing tree growing practices in different parts of the world
it is apparent that trees are often most prominent in situations
where labour, capital and physical resources are limited. In
such situations, trees can play one or more of the following
overlapping roles:

* trees can help maintain productivity of land in situations
  of scarce capital, and can substitute to some extent for
  purchased inputs of fertiliser and herbicides and investment
  in soil and crop protection;

* in situations where capital and labour is scarce, trees -
  because of their low input and management requirements - may
  be the most effective use of these resources;

* trees may provide the best income earning opportunities when
  the size of landholdings or the productivity of land falls
  below the level at which the household's basic food needs
  can be met from on-farm production of food;

* trees may enable farmers to spread their risk by
  diversifying their farm outputs, evening out the seasonal
  spread of inputs and outputs, and building up a stock of
  capital in the form of mature trees that can be harvested
  and sold for cash during emergencies.

4.6.5   Tree Cash Crops and Household Food Security

In principle, an increase in a household's income should improve
their access to food. In practice, however, the shift from
subsistence to cash crop production has in some cases lead to

reduced household food security, with adverse affects on both the stability and quality of food supplies and the nutritional well--being of children. Increased food prices, loss of employment opportunities, vulnerability to fluctuations in cash crop prices, fluctuations in the availability and prices of marketed foods, and the reduced control that women have over household resources are some of the factors that have been identified as contributing to this (Longhurst, 1987).

Potentially, therefore, tree crops could have a negative impact on household food security. Tree growing can transfer land from food crop production with a loss of employment; tree promotion services are concentrated on male farmers, often there is only one marketable product with few market outlets; and trees take several years to mature.

In practice many of these potential negative impacts are offset by other features of tree growing. As was noted earlier, the transfer of land from food to cash crops is often in response to changing conditions which make food crop cultivation impracticable (e.g. increasing scarcity of land or labour). Tree growing can provide farmers a means of keeping their land in productive use with minimum labour inputs.

The impact of tree growing on household food security depends on the types of trees grown and the way they are managed. If land previously used by women to grow subsistence food crops is converted to eucalyptus pole plantations controlled by their husbands, then trees may have a variety of detrimental effects on family food security. On the other hand, most farm tree species provide products such as fodder, food, fuelwood, mulch, shade and soil protection in addition to generating cash income.

There is a danger, however, that forestry programmes which encourage tree planting on farms could induce farmers for whom it is not appropriate to shift to tree monocrops. Cash incentives and a concentration on a few species familiar to foresters but unsuited to household needs could have a negative impact on a household's food security. These dangers may be exasperated by the pressures to achieve the ambitious targets that are a feature of many large "farm forestry" programmes.

4.6.6   Trees as a Form of Insurance

Vulnerability to emergencies and other contingencies, and the inability to provide for them, are important and often neglected aspects of poverty. Emergencies such as sickness of a family member or loss of assets through theft, fire or flood are by their nature unpredictable. Large periodic expenses such as weddings are more easily foreseen. In either case they can present a major drain on a family's resources, requiring assets to be sold or mortgaged, or cash to be borrowed - often at exorbitant interest rates. For a family that is poor already, such events can push them further into poverty, seriously undermining their ability to obtain food and other basic necessities.

Trees can provide a useful way of coping with contingencies. In
many parts of the world they are used as a form of savings that
can be drawn upon to meet such needs. In some cases they are
deliberately planted with this in mind, to be cut for timber or
fuelwood when large cash needs arise.

As a form of savings, trees have a number of advantages. They
require very little investment of capital, unlike other methods
of savings such as livestock or paddy land. Under favourable
growing conditions, their value appreciates and they are not
too susceptible to inflation. They can be harvested when needed,
and in the amount required - and some trees coppice when cut, so
that the investment will re-establish itself for minimal extra
cost.

Growing trees, of course, is not without risk. They have to be
protected from damage by animals and fire. Marketing may also
pose problems, especially for poor farmers with only small
amounts to sell. In some cases, rights of tree ownership are
ambiguous, or farmers may have to go through a lengthy process
to get permission to cut trees. Trees, therefore, may not be
an ideal form of savings, nor one which is open to everyone. But
for many rural families, it represents a cheap and practical way
of providing for contingencies (Chambers and Leach, 1987).

4.7 Land tenure and food security

Underlying many aspects of forestry and food security is the question of land tenure. Who owns land - and who controls it - has a crucial bearing on who can benefit from crop lands, trees and forests, and who cannot.

4.7.1   Distribution of Land-holdings

Because control of land is such an important and sensitive issue, obtaining information on its ownership is seldom an easy task (Chambers, 1983). Although accurate data may be hard to obtain, the general pattern of land holding in most Third World countries is clear.   With a few notable exceptions, the distribution of land is highly unequal. The available figures vary from country to country, but it is not uncommon for less than 10 percent of the landholders to control more than 40 percent of the total arable land.   In some countries, particularly in Latin America and Asia, the concentration of land in the hands of the rich is far greater.

Gradations of poverty exist even in the poorest of communities, where families with small plots of land are substantially better off than those with little or none (Castro <u>et al</u>, 1981).   Many households lack even permanent rights to the lot on which their house stands (Herring, 1983).

4.7.2   Ownership of Trees

It is important to distinguish, however, between land tenure and tree tenure, since the two are often different. In many cases, ownership of land does not grant automatic rights to the trees growing upon it. (Fortmann and Riddell, 1984).

Large timber trees in Central Kenya, for example, are often regarded as the property of extended kinship groups even though they might be situated on essentially privately-held land (Castro, 1983).   In Papua New Guinea it has been noted that an individual can obtain a proprietary interest in trees of economic importance such as coffee, pandanus nut, and highland betel nut, by planting them, or by inheritance or gift. Having such an interest does not in itself confer rights to the ground below the trees.   Thus, one may receive as a gift a grove of pandanus trees, but the land upon which they grow remains the property of the grantor or his kin group (Grossman, 1984).

In other instances, tree growing instills the rights to the land on which they grow (a common practice throughout humid West Africa). For this reason few farmers are allowed to plant trees on the lands they farm which often "belongs" to Chiefs or extended kinship groups (Gastellu 1980 in Falconer, 1989b).

Rights to woodland are also sometimes defined differently from those governing access to farmland (Fortmann and Riddell, 1984).

Even in areas where farmland is normally private, woodland may remain under the jurisdiction of communities or other local groups. In Nepal, measures have recently been introduced to return areas of forest that were previously nationalised to village control.

In some countries, such as the Dominican Republic and Honduras, the ownership of all the trees in the country is officially vested in the State. There are penalties for cutting trees without permission, even those standing on a farmer's own land. Although designed to protect trees, this kind of legislation often has the opposite effect and discourages farmers from taking the initiative and planting trees themselves (Murray, 1981).

Finally, it should be noted that the rights to exploit different forest or tree products (whether on farmlands or in forests) often vary from those associated with tree ownership. For example, people may have the rights to collect medicines and foods from forest trees but not to sell the tree for timber or fuelwood. Often, traditional tenure practices provide fairly open access to subsistence forest goods (e.g. foods and medicines) while those with commercial or symbolic value may be more restricted (Boamoah 1986 in Falconer, 1989b).

Tree tenure systems have profound influence in determining the role forests and farm trees can play in household food security: as it is often a crucial factor in inhibiting or stimulating tree growing (Fortmann, 1984). In some cases, these systems may evolve as changes in the rural economic as well as physical environment will change the value of different tree and forest products.

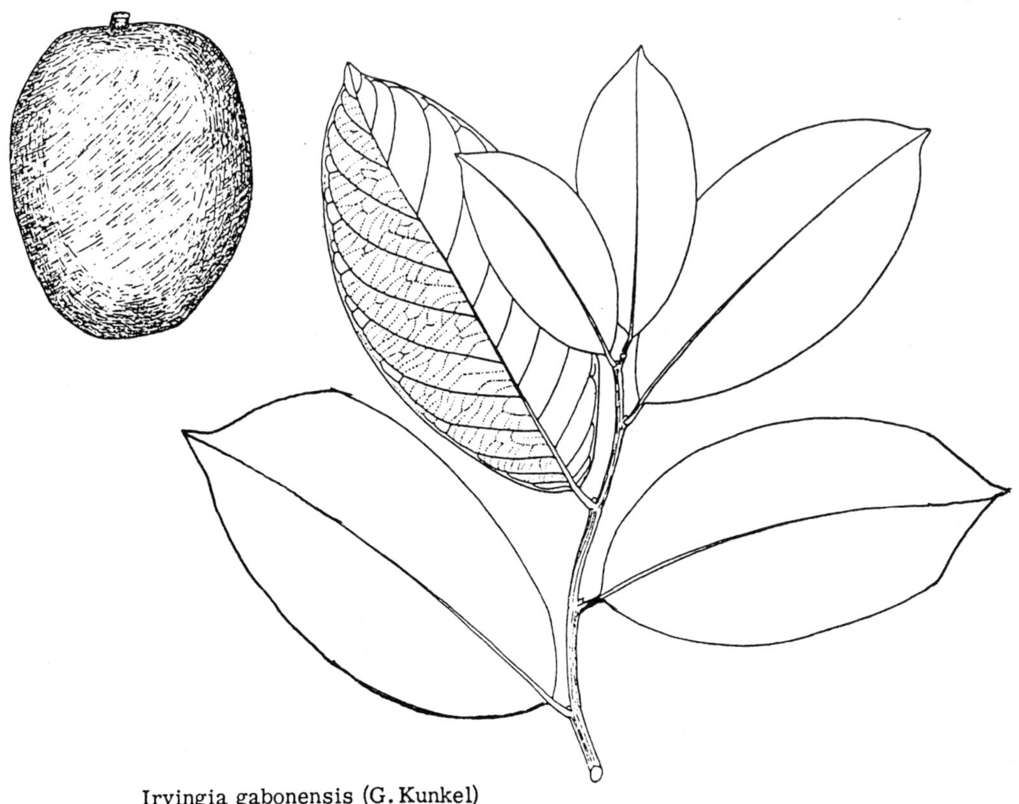

Irvingia gabonensis (G. Kunkel)

4.8 Common property resources: assuring household food security

In many parts of the world, especially in Africa, substantial areas of forests and woodlands remain under various forms of communal control. Access to the food and other forest products they provide is determined by traditional rules and customs, backed up in some cases - though not all - by formal legislation.

These common property resources are in many cases coming under increasing strain as a result of growing human and livestock populations, nationalization of forest and rangelands, increased privatisation of these lands, and a variety of other forces. How these strains are accommodated has a major bearing on the welfare and food security of the many families who depend on them.

There is a widely held notion that in the wake of increasing population pressures privatisation is the only way of protecting common property resources from over-exploitation (Hardin, 1968). Resource management systems based on communal rights are often thought to be inherently inefficient and lead to deterioration of natural resources - as each individual seeks to maximise gain. Behind this notion lies the assumption that in all common property systems everyone has open and unrestricted access to the resource. It is wrong, and indeed misleading, to assume that this is the only and general way common property resources are and can be managed (Dani et al, 1987). Many traditional common property systems are being ignored. Others are being replaced by privatisation on the questionable assumption that this will be a more effective basis for management.

4.8.1 Diversity of Common Property Systems

There are, in fact, many different types of common property systems. The majority incorporate mechanisms to protect them against abuse and over-exploitation.

Pastoralists are a case in point. Usually, they have highly developed systems of rangeland management, with mutually recognised rights and duties. The Maasai, for example, traditionally had "elaborate grazing sequences, grazing flushes to create hay in dry-season reserves; regular use of donkeys to carry water... to permit camps to stay away from their dry season reserves as long as possible... and regular social rebuke and avoidance of families or camps that fail to adhere to good management practices", according to one description (Jacobs, 1980). This is no free-for-all, but a carefully regulated system involving defined and enforceable rights and duties, evolved to meet both social and environmental needs.

In many parts of the world smallholders still retain group or corporate land tenure of various forms (Erasmus, 1977). The distinctive feature of this is that rights to land are ultimately vested in a local social group such as an extended family, caste, tribe, or village. A member of the group has inheritable rights to use the land within the jurisdiction of the community, but no rights to sell it.

One of the most important features of these systems in terms of
household food security is that they are locally controlled and
thus flexible-- in the case of forest resource management this
recognizes the fact that they are most essential during emergency
and seasonal hardship periods.

4.8.2   Externally-Imposed Common Property Systems

There has been little success with externally-imposed common
property systems as they have been planned by outsiders who have
rarely been able to comprehend local conditions sufficiently well
to be able to devise a system geared to local needs, values and
aspirations.

Group ranches, for example, were for a decade or more a popular
and expensive form of development aid to African pastoralists.
The success rate in these projects was extremely poor; some
observers claim that not one thriving group ranch can be found
in contemporary Africa (Dyson-Hudson, 1985).

Community woodlots have also been widely promoted by development
assistance agencies.   They were based on assumptions that
communities would cheerfully and effectively co-operate in
planting trees, caring for them and protecting them, and that an
equitable distribution of benefits could be assured. Generally,
however, the results in India, Africa and elsewhere have been
disappointing.

The heterogeneity of communities, the differing interests of
their members, the scarcity of land and uncertainty about tenure
status, the problems in the distribution of benefits, and the
lack of a general structure of cooperation, have all been
contributory factors.   The problems is that "the close
interdependence of members required by community schemes cannot
be fostered by decree" (Cernea, 1985).

4.8.3   <u>Building on Existing Institutions</u>

There is no doubt that, in the past, common property management systems were a widespread and effective means of managing many natural resources, including forests, rangeland, water, and fisheries. In many communities, because of population growth, market forces, privatisation, state interventions and other socio-economic changes, the rules have broken down and these traditional systems have been impaired. Yet in spite of this, many of these systems still play an important role in the management of scarce natural resources, complementing and combining with systems of private rights (Runge, 1986).

By building upon the existing institutions, it may be possible for local people and external agencies to co-operate in devising common property management systems that take into account local factors, allow forest access on certain conditions to poorer people, and ensure conservation of the natural resource. No one approach will be appropriate for all situations. Much depends on local circumstances, on the traditions that exist of collective action, and on the quality of local leadership.

The point is, however, that for many rural families, especially poor families, common property resources are the only resources available to them. In order to promote food security for these groups far more attention will have to be given to the effective management of common property resources.

Chapter 5 Opportunities for action

There are no simple prescriptions to follow on how to integrate food security objectives into forestry activities. Experience is still limited and there are few examples of forestry initiatives - successful or otherwise - that have been designed with food security as a specific target. Nonetheless, there are many opportunities for action, some of which are sketched out briefly in this chapter. They provide a starting point for further discussions on how forestry can better contribute to food security, and a basis on which practical actions can be devised. Support at the policy level will be essential.

5.1 Setting the policy framework: diversifying forestry activities to meet people's needs

For forestry to play a role in enhancing food security a broader and more flexible policy framework is needed to provide the necessary support for specific programmes and initiatives.

5.1.1 Defining Policy Objectives

National forest policies are usually expressed as general statements of intent which encompass a wide range of objectives, including production, environmental and developmental goals. In the past, forest policies have tended to be directed primarily towards maximising revenues and foreign exchange earnings from the forest, and ensuring supplies of raw materials to large forest-based industries. As a result, the needs of local people have often been relegated to second place.

Focusing on food security requires a fundamental shift in mandate in _who_ forests are managed for: changing from timber merchants and government treasuries to emphasis on local people. In so doing, a much broader and more flexible set of policy objectives are needed allowing for a diversity of programme options. This will involve incorporating the food and income needs of local people as a much more prominent element in overall forest policy, and expanding forest planning and management practices to include greater involvement of local people.

In addition, a review of existing forest policies is needed in light of food security concerns. Of particular importance are:

* changing policy measures which prevent the desired broadening of forest uses (for example, changing legislation that discriminates against users of non-timber forest products or access to wood for small-scale enterprises);

* replacing legal and other constraints that discourage tree growing outside forests with incentives and other measures that promote more effective use of trees within farming systems;

* developing and enforcing regulations to minimise the negative impact of large wood-using industries on the local environment and local people;

* making modifications in forest laws to recognise the needs of landless and poor families, and to extend their involvement in food and income generating activities based on the forest.

5.1.2   <u>Land-Use Policies: Promoting Sustainable Uses</u>

Land-use planning and policies and land tenure issues are all of
central importance to food security problems. As was discussed
in Chapter 2, many environmental problems caused by forestry
activities such as timber harvesting on unstable slopes (e.g.
resulting in land slides, river and irrigation channel siltation)
can have disastrous effects on food production and supply. Issues
of control and management of forest areas, as well as people's
rights to forest products on their own farms all have important
implications for the ways in which forest and tree resources can
be used to help counter food security problems.

<u>Land-Use Policies</u>

In some cases land-use policies of different divisions within a
government promote conflicting goals. In addition, government
policies may encourage, either actively or tacitly, practices
that are environmentally unsound and are, in development terms,
disastrous for the poor.

The clearance of fragile tropical forests areas to open up
grazing lands for cattle ranching or unsustainable grain
production is one of the most obvious examples. All too often
this has led to rapid loss of soil fertility, followed by erosion
and permanent land damage. Although it may boost food production
in the short term, its long-term effects are to degrade the
resource base and to destroy any potential for sustainable
production.

<u>Land reform</u>

On the question of conversion of forest land to agriculture, most
agrarian reform programmes and forest settlement schemes have
considered forests as a land bank for agricultural expansion.
In many cases, there has been inadequate consideration of the
suitability of this land for sustainable food production. As a
result, there are large areas which have been degraded to the
point of minimal productivity, despite huge investments in land
clearing, infrastructures, subsidies and incentives.

A more careful and creative approach is needed, both in selecting
land for conversion to agriculture, and in encouraging more
sustainable use of the land perhaps through combinations of trees
and crops. In some cases there are alternatives to land
clearance that may be more effective in providing settlers with
a sustainable livelihood. These involve retaining forest cover
and developing the land for forestry uses, whilst promoting
employment and income earning opportunities. Increasing the local
opportunities for forest exploitation could provide people with
a sustainable living without running the risk of environmental
degradation.

It is important that land-use planning is not treated as merely
a scientific exercise, divorced from the realities of local
circumstances. It is only by incorporating local needs,
perspectives and know-how that realistic plans can be developed
- plans that take into account the constraints people face and
the often difficult conflicts of interest involved.

Control of Forest Land

When it comes to the use and control of forest land, the custodial approach is still the dominant forest policy. Retaining custody of forest lands, however, should not be a goal in itself. While there may be strong arguments for maintaining control of forest land in some situations there is a need for alternative management approaches. In many instances the food security needs of local people can be better served with flexible and more closely monitored forest management which responds to local needs and circumstances. These may involve the partial or complete transfer of forest land to local ownership, and the devolution, to varying degrees, of responsibility for control and use of forest resources to local communities.

Shifting cultivators clearing forest land to plant their crops in Thailand

In Thailand a 'forest village' system has been designed to settle shifting cultivators in forest areas. Villagers are employed by the forest department to establish plantations, and are allowed to plant crops between the young trees. Each family is also allocated 2.4 hectares of farmland for their own uses. No land title is issued, but user permits are given which can be inherited but cannot be sold. In addition, the forest department assists in providing housing, vocational training, schools, and other infrastructural support.

Developing alternative management and custody schemes requires detailed knowledge of local circumstances, and it depends both on winning the trust of local people, and on providing them with the support they need to ensure sustainable management of the forest resource. In the long term, the benefits of developing local capacities to tackle problems through joint action may be even more important than the short-term gains from the tree products or services provided.

5.1.3   Holistic Approach

As was noted at the outset food security is a complex issue as are the rural environments in which foresters operate. Thus a more holistic approach at all levels is needed including planning, research, programme development and forest management. An approach which will draw on and integrate the experiences of many people and institutions. As was noted above, co-ordination between forest policies and those in other sectors of the economy is important for countering food security problems. There is an urgent need to shift away from the current, narrow sectoral approach of policy making, to one in which policies in forestry, agriculture, livestock, industry and development are integrated so that they complement rather than compete with each other. In forestry programmes geared to improving the welfare of forest dwellers, for example, substantial inputs of resources from other non- forestry sectors such as education, health, and infrastructures, will be needed. Integrated programmes have far greater potential for tackling food security problems than forestry initiatives on their own.

The linkages between forestry and agriculture deserves special attention. As was shown in Chapters 3 and 4 trees grown in agricultural lands contribute to household food security in many ways. This contribution has been largely ignored, both by foresters and agriculturalists.   There is a clear need to strengthen co-operation between foresters and agriculturalists in planning extension and programme implementation, so as to create more effective partnerships for working with farmers and the problems they face.

5.2 Institutions: support for food security objectives

Changes in policies alone will not change forestry activities to
include food security objectives. Foresters and their
institutions will.

Present institutional structures are in many cases poorly matched
to the challenges set by forestry and food security problems.
What is required is much greater co-operation between the
agriculture, forestry, livestock and other departments involved
in development and natural resource management. In addition,
within Forestry institutions themselves a broader base of
professionals are needed. Notably, more staff with training in
social sciences. New types of training will be required for
forestry professionals and extension workers to provide them with
the skills needed to work more closely with villagers.

There is no doubt that important strides have been made in recent
years in a number of countries in broadening the role of forestry
to meeting people's needs. For those involved, one of the
problems is that career and reward systems do not always
favourably regard such activities. There is a high risk of
failure and error when doing innovative projects. These problems
highlight the fact that adequate institutional support is
required for new programmes, especially for those geared to
working at a community level.

Foresters, especially those at the lower levels, who are in
contact with local people, will need to develop new roles:
becoming brokers between central government and local people;
interpreters of local needs and aspirations; facilitators,
helping local people to organise and play an active part in
forest management; and advocates for the poor and disadvantaged.
Such roles represent radical departures for most forestry
services, but they are necessary if forestry is to address food
security problems. They also represent a radical departure from
most extension service approaches which seek to present
predesigned solutions. Two-way development communications are
needed with local input into the extension programme design.

A variety of new organizational approaches are also open to
forest departments. One possibility is to create a formal
separation of the police and extension functions of forest
departments, as has happened recently in Senegal. In Nepal, some
villages have been encouraged to set up "village forest police"
to assist in the protection of local forests. The aim in both
cases has been to create a partnership, rather than an
adversarial relationship between the forest service and the
people (Lai and Khan, 1986).

The need to include more women in research, planning and
extension agencies is also crucial. If efforts to increase food
security are to be successful, women and women's concerns must
be systematically incorporated into institutions and programmes.
Employing more female staff, at all levels, is one of the best
ways this can be achieved.

Finally, Forestry Departments can use the expertise of other institutions - especially those which are already promoting community level development. Prominent among them are non-governmental organisations (NGOs). NGOs can be particularly useful, innovative and flexible; and they are often well suited to working at the community level.

5.3 Research priorities

Throughout this report the tentative nature of our understanding of the links between forestry and food security have been stressed: from the effects of deforestation on rainfall levels to the importance of forest foods during emergency droughts. There is little doubt that there is a great need for research on all aspects of forestry and food security.

Of greatest importance is research which addresses the key issue of how foresters can integrate food security objectives into their activities. In general, forestry initiatives will center around four main priority areas: maintaining and developing forest products crucial for local food security (e.g. supply of forest foods); minimizing the negative impact of forestry activities on food security (e.g. harvesting practices and water quality degradation); improving food production (e.g. improved agroforestry and conservation techniques); and increasing the returns of forests for local people (e.g. development of small-scale processing enterprises). Also imperative is a better understanding of how people use and manage their surrounding forests, especially in terms of their food security needs. Building on this knowledge base will provide one of the greatest research opportunities.

There is a wealth of knowledge in many rural communities about forest species, their ecology, their management and their uses. In some areas, however, it is disappearing as the natural resource base changes, societies change and traditional practices die out. In these instances, research is needed urgently as this information will provide invaluable clues about the ecology of different species, about how they can be managed for sustainable production, and about their uses.

At the technical level, there is a whole range of possibilities for improved management of forests and better use of trees on the farm. Some of the most important research priorities can be summarised as follows:

* research to develop techniques for reclaiming denuded and degraded areas - for example, land affected by salinisation or desertification - using trees and agroforestry methods, with the aim of returning these areas to productive uses;

* studies to understand the effect of forests and trees on moisture availability for agriculture - this will involve such issues as the effect of forests on distribution of rainfall, groundwater recharge, and flooding. Studies are needed both at the level of individual watersheds, and for major river and mountain systems such as the Himalayas, the Nile system, and the Amazon basin;

* investigations to elucidate the mechanisms of plant-to-plant and plant-to-animal interactions in mixed land-use systems, and to evaluate the complementary and competitive interactions among different components of these systems, and to suggest ways of optimising sustainable production;

* research to evolve low-cost alternatives to minimise the
  requirement for external inputs of fertiliser and
  pesticides, and maximise the benefits from nitrogen
  fixation, nutrient cycling and organic matter addition in
  tree and crop mixtures;

* studies to increase sustainable production from forests and
  agroforestry systems through species selection, improvement
  of the genetic stock, and new breeding and propagation
  techniques;

* research into the sustainable management of trees and
  forests for multi-purpose uses;

* research aimed at the identification, management and
  enhanced use of under-exploited species of plants and
  animals in forest habitats - this should be co-ordinated
  with efforts to conserve genetic resources using both
  <u>in-situ</u> and <u>ex-situ</u> approaches.

As with other aspects of the development of programmes aimed at
food security objectives, researchers need to broaden their
methods - undertaking interdisciplinary studies and devising
methodologies which are urgently required if multi-component
forest and farm systems are to be developed. These need to
incorporate both biological and socio-economic aspects of
production systems.

An underlying priority, however, is to ensure that research
activities, whether carried out by universities or other
institutions, are firmly grounded in the realities of local
circumstances and local problems. Too often, research is carried
out for its own sake, and becomes divorced from the development
process it is supposed to be serving. An increased emphasis
on-farm research is one way to guard against this tendency, and
will in many cases be essential if new techniques are to be
successfully transferred from the laboratory or research
institute to the farm.

Finally, there is great need and scope for research on the
methods and institutional arrangements required for instituting
forestry programmes with food security objectives. What are the
different types of flexible management practices, for example,
with which forestry departments can experiment? What types of
strategies and approaches will effectively address local peoples
needs? It is imperative that researchers do not focus only on the
products and services of forests and trees, but also address the
social and economic conditions needed so that people can benefit
from them.

5.4 Approaches

There are many ways of diversifying forestry activities so as to address at least some food security problems. The following discussion briefly sketches some possibilities for action; it should not, however, be seen as directive - it is intended to illustrate how many forestry activities could be adapted and developed to help counter local food security problems.

Of utmost importance in terms of developing programmes is identifying the nutrition problems in a particular area, and the people the programme is intended to benefit.

5.4.1.   Identifying the Problems

If food security considerations are to guide the design of forestry projects and programmes, an understanding of the nutrition problems and shortfalls in local diets is needed. For example, in some areas specific nutrients are lacking from the diet (e.g. niacin in areas with maize-based diets) - or simply there is not enough food during one season. In the first instance, introduction of farm trees with niacin rich foods could prove beneficial. In the second instance, the cause of the food problem will determine the possible solutions. If the problem is shortage of food supply, one may consider introducing food bearing farm trees or facilitating access to forests during the hunger season. If the problem is lack of income to buy food, development of forest-based processing enterprises might be useful. It should be noted here that it is equally important for a forestry project or programme to decide which problems they cannot address as to decide which ones they can.

In identifying dietary problems some basic information will be needed (often available from country health or nutrition departments): an idea of the important components of the diet and any widespread nutritional deficiencies (e.g. vitamin A deficiency), identifying who within a community is particularly susceptible to food problems, an idea of the seasonal variations in the diet, and the recourses people take in times of emergency, information on the availability of food in markets, and the extent to which people depend on purchased foods. In addition, information on the general "socio-economic climate", e.g. the income earning opportunities of the area will be required.

5.4.2 Identifying Target Groups

As has been stressed throughout this section, gearing forestry programmes to food security objectives means gearing activities to the needs (or at least the concerns) of local people. "Local people" are of course not a homogeneous group waiting to be identified. There are many divisions within communities, which may well be at the root of some of the communities food security problems. These factors, however, do not detract from the fact that programmes and projects geared to local food security problems must be planned and managed at the local level.

A great many factors will determine who a programme is being
designed for (and by); and these will depend entirely on the
particular local conditions. There are nonetheless a few issues
to consider at the more general level, notably:

* nutritionally vulnerable groups within a community and the
  groups or families most dependent on forest or tree
  resources for their well-being;

* the central role of women in food production and food
  security.

Finding out which groups are particularly vulnerable to food
security problems is essential if efforts are to be concentrated
on those most in need. This will vary from region to region.
Some of the groups which are often most at risk include:

* the landless poor - who depend on wage labour for income
  and who often rely heavily on dwindling common property
  resources for firewood, fodder and other basic needs;

* forest dwellers and shifting cultivators - who often suffer
  due to their lack of secure land tenure and the increasing
  outside pressures on forest resources and forest land;

* small farmers who lack the land and resources needed to
  guarantee adequate subsistence production or income
  generation - many of whom are subject to the combined
  threats of environmental degradation, declining fertility
  and continued fragmentation of landholdings;

* pastoralists and herders - especially those in fragile
  environments that are susceptible to droughts, and in areas
  where rangelands have been reduced through encroachment by
  cultivators, exclusion by government, and other factors;

* young children have particular dietary needs which cannot
  be met in especially poorer households.

Although the reasons behind their food security problems differ,
these groups do have a number of features in common; frequently
they lack an effective political voice, they are short of
capital and other resources, and they are marginalised from the
mainstream economy - and the health, education, economic and
other benefits it provides.

What these groups do not lack, however, is ingenuity and the will
to better their own lives and those of their families if they
are given the chance. The challenge is to devise programmes and
approaches that are relevant to their needs, and provide them
with an opportunity to improve their own income and food security
position.

5.4.3   <u>Importance of Women</u>

Given the central role that women play in food production and
food security, involving women, and their concerns will provide
invaluable insight for programme planning and direction. The
possibilities for successfully integrating women into the project

planning process would be enhanced if women staff and extension agents are used.

The demands on women's time may be of central importance to designing programmes from which women can benefit. If women are to be involved, for example, in a tree growing project, the jobs of weeding, watering and other tasks will have to be fitted into an already busy work schedule. Projects will fail if the demands they place on women's time cannot be accommodated, and if the benefits are not seen as being worth the extra work required.

5.5 Important lines of action

<u>5.5.1</u>   <u>Diversifying Forest Management to Incorporate Locally Valued Products</u>

From the point of view of the food security of those living in and around forests, there are a number of important ways in which forest management can be improved: by focusing on non-timber forest products so as to provide a wider and more plentiful range of products; by incorporating products of local importance in plantation development, and by providing better and fairer access to the resources that already exist.

The expression 'minor forest products' sums up well the perception of non-timber forest products within conventional forestry circles. They tend to be treated as peripheral; an added bonus that may be of some interest to local people, but not something that is a major concern for the forestry authorities.

Forest management needs to focus on improving existing forest resources - especially those of local importance . New skills will be needed which focus on management for multiple products and forest uses. This does not mean that traditional production goals have to be abandoned. What is needed are techniques that combine these goals with the supply of other products of use to local people - products such as bushmeat, rattan, bamboo, firewood, traditional medicines, fruits, honey and other forest foods.

In some cases this may involve taking measures to conserve useful species or particular areas of natural forest, instead of clear felling. Alternatively, it may require deliberate efforts to protect or introduce certain desirable species. Either way, research will often be needed as a preliminary step. The traditional focus on high-value timber species has meant that experience in how to manage and harvest the many other plants growing in the forest is often limited. Knowledge on sustainable use of forest wildlife is also poor, and needs to be developed if their potential as a food source is to be sustained. A first step will be to build from the local knowledge of the forest ecosystem and management practices. This will provide foresters with vital information, and will pave the way for more cooperative management - including local people.

Forest and tree crop plantations are generally believed to destroy non-timber forest product supplies, most notably wild animal habitat. While this is true of products which are cleared to make way for the plantation, there is scope for incorporating locally valued products into plantation management. Simple management techniques such as leaving patches of forest vegetation may help sustain wild animal populations. "Intercropping" plantations with sought after products could also be undertaken; or creating hedgerows of mixed vegetation to encourage desired animal species. In some cases management for non-timber forest products could provide a steady (as opposed to lump-sum) source of revenue to offset the cost of management.

Improved production is one side of the coin; the other is better
access. If forest products are to benefit local people they have
to be available, and there have to be management mechanisms in
place that ensure that they can be exploited in a sustainable
fashion.

Again, new approaches will often be needed, and these will have
to be worked out on the basis of local conditions. In some
situations the most effective approach may be for the forest
department to retain close control over access to resources,
using permits, for example, or by managing the harvesting of non-
timber products and distributing them direct to local users.

Elsewhere, however, other more innovative approaches might be
more appropriate, and more sustainable in the long-term. Certain
areas of forest might be set aside specifically for local users
under a variety of stewardship agreements, either with
individuals or communities. Special arrangements could be made
to allow women access to forest resources, or give concessions
to disadvantaged groups. In this way, it may be possible for
benefits to be targeted to those in greatest need.

In the case of forest dwellers who have been living in forest
areas for many years, securing their rights of ownership and
access to forest resources may be one of the most effective ways

of encouraging better management. Often they have extensive
knowledge of the local ecology but have been prevented from using
it by their insecure tenure or because of interference from
outside. Providing these groups a direct stake in the continued
management of the forest, within the framework of clear ownership
and access regulations, could bring major benefits. Instead of
being part of the problem, as forest dwellers have often been
regarded, they could become part of the solution.

Effective alternative forest management schemes cannot be
expected to emerge overnight. Inevitably, there will be failures
as well as successes. With imagination and commitment, however,
they do offer genuine opportunities for giving local people
responsibility for the protection and management of forest
resources. In the long-run this may prove much more realistic
than the conventional policing approach to forest management.

5.5.2   Encouraging Tree Growing on Farms

One of the most promising approaches to increasing the food
security of families with land is to encourage tree growing on
farm and fallow lands. As has been discussed this can contribute
to food security in a number of ways; by providing food and
animal fodder directly, by improving the conditions for crop
growing and livestock rearing, and by supplying products that can
be sold for cash.

As a broad approach to sustainable agriculture, agroforestry
systems have undoubted potential, especially for poor farmers who
cannot afford to use fertilisers and other external inputs. The
traditional systems that already exist in many parts of the world
can be enhanced and diffused. There is also enormous scope for
developing new and improved forms of agroforestry, using new
species combinations, better genetic stock, and a range of new
techniques. To put this potential in context, however, a number
of points need to be recognised:

* the needs and perspectives of local farmers must provide
  the guiding influence in designing and optimising
  agroforestry systems. Their relative need for fodder, food,
  wood products, income, and other benefits will dictate to
  a large extent the most appropriate system;

* on-farm research involving farmers themselves is critical
  in taking new agroforestry techniques out of the research
  stations and into widespread use;

* many of the detailed interactions between trees and other
  components in agroforestry systems are still only partly
  understood, and most of the potential species combinations
  and management approaches have yet to be properly assessed;

* trees can compete with, as well as enhance, the production
  of crops. The tree component within agroforestry systems
  therefore has to be carefully designed;

* the technical options available are highly dependent on the
  agro-climatic conditions. Techniques that work well in

humid regions can rarely be transferred to arid and
semi-arid areas without considerable modification and
adjustment;

* new integrated management approaches need to be matched to
  local market opportunities, as well as to agro-climatic
  conditions;

* agroforestry is not the solution everywhere. There are
  many cases where existing farming and livestock systems are
  performing perfectly well, and there will be little to be
  gained by introducing more trees.

Thus, while agroforestry approaches offer exciting opportunities
for improving rural livelihoods and enhancing food security, they
have to be firmly grounded in local realities and be tested under
local conditions.

5.5.3   <u>Supporting Small-Scale Forest-Based Enterprises</u>

Huge numbers of people already depend on gathering and processing
tree and forest products as a source of income. These include
products grown on the farm and those obtained from the forest.
By supporting these activities, and helping to make them more
profitable and sustainable, the livelihoods of those concerned
can be improved and their food security enhanced. This is of
particular relevance for the landless and other disadvantaged
groups, as these are the people who generally depend on these
activities the most. Women specifically stand to benefit.

A number of possible options can be identified:

* guaranteeing supplies of input materials from government
  forests, at controlled or reduced prices, and making sure
  small-scale enterprises are not subjected to unfair
  competition from larger industries;

* increasing the 'value added' from tree products by
  supporting more extensive processing by local people;

* strengthening the managerial and business capacity of small-
  scale enterprises by encouraging producer cooperatives,
  associations, and other joint groups;

* developing and promoting new technologies that improve the
  returns, improving efficiency or product quality;

* providing tax and other incentives to encourage the
  establishment of small-scale enterprises;

* improving the availability of credit to small-scale
  enterprises to allow them to expand their capacity, create
  more employment and increase turnover and profits.

Of course, the development potential of different small-scale
enterprises will depend on a number of local conditions, notably
the raw material supply, the market potential, market access and

supply of labour. More information will be needed to understand
how to identify enterprises with long term viability from those
which will shortly be overcome by larger operations or substitute
products. Information will also be needed in order to better
understand how such enterprises may be supported in a way that
benefits accrue to the poor.

5.5.4   Providing Market Support

For those involved in the sale of tree-based products, either
from the farm or the forest, the benefits they obtain are
directly linked to their access to markets .   In many cases,
local collectors and processors receive very little for the
products they sell.   Instead, most of the benefits are captured
by middlemen and urban traders operating further along the
marketing chain.

There are a variety of measures that can be considered as a
possible way of assisting in the marketing of forest products,
with the aim of improving rural incomes:

   * strengthening the bargaining power of producers by setting
     up marketing cooperatives, or producers' associations;

   * providing farmers with better market information to raise
     awareness of market opportunities and limitations, warn
     them of possible fluctuations in market prices, and assist
     them in diversifying what they produce so as to reduce
     risks;

   * supporting the marketing of tree products by providing
     transportation and storage facilities, linking sellers with
     buyers at markets and fairs, and giving advice on
     advertising and marketing strategies;

* assisting women in the marketing of tree products by
  ensuring they have direct access to market outlets and
  receive personally the proceeds of the goods they sell;

* mounting promotional campaigns to encourage consumers to
  buy indigenous tree-based products instead of imported
  alternatives;

* reviewing price controls that set a ceiling on the price of
  tree products and discourage sustainable production.

Interfering with market forces is always a delicate business.
The side-effects are often hard to predict and it is not unusual
to end up having the exact opposite effect from that originally
intended. Measures to support rural producers by setting minimum
price levels, for example, may result in reduced consumer demand
and a shift to alternative products - thus negating the supposed
benefits. Subsidies also have to be used with discretion;
besides being expensive and difficult to administer, they may
foster an unhealthy degree of dependency amongst intended
beneficiaries and become very difficult to remove once they have
been introduced.

To be effective, government interventions into market systems
need to be carefully researched and properly targeted. Where
there is a case for providing subsidies and other forms of direct
support it is often best if these measures are introduced for a
clear and limited time period, and are phased out once they have
had their desired effect. Similarly, rather than government
agencies continuing to supply market information and other
services, it will often be more efficient if these
responsibilities are passed over to producer groups themselves,
which once established may be better able to provide the
necessary ongoing services and support.

5.6 Concluding remarks

While forestry efforts alone cannot substantially alter social, economic and political factors at the root of many food supply inequalities, they can support and build from the contributions forests (and farm trees) make to household food security. In order to strengthen and develop these contributions, foresters need to focus on new goals and devise new approaches for their activities. This will entail modifying existing institutional approaches and arrangements, and the traditional focus of forestry training, research and extension work, as these are not well matched to the task of addressing food security objectives.

Food security issues are especially important at the policy level. Supportive policies must influence the direction of programmes and projects in order to optimise their impact on food security and rural development. These issues are complex relecting the ever-changing rural world, especially for the poor in terms of their access to physical, capital and labour resources which must be juggled in order to survive and develop.

The preceding chapters have shed light on some of the links between forests, forestry activities and people's well-being - their year-round food supply. This focus on food security highlights the fact that forests (and thus forestry activities) cannot be isolated from their rural environments; they are intricately linked with the physical and socio-economic factors which sustain people living in, or near them. On a grander scale forests may also be linked to the "world" environment affecting climate patterns and thus the lives of everyone. Although foresters may feel that food security concerns are far beyond the bounds of their profession, their activities directly affect the food security of households in their country or region.

Annex 1

List of documents presented at the Expert Consultation on Forestry
and Food Production/Security, Trivandrum and Bangalore, India,
8-20 Feb. 1988.

Arnold, J.E.M. Tree Cultivation, the Household Economy and Food
Security. The Oxford Forestry Institute. Satellite Paper.

Arnold, J.E.M. and Falconer, J. The Socio-Economic Dimensions of
Forestry and Food Security. The Oxford Forestry Institute. Main
Paper.

Arnold, J.E.M. and Falconer, J. Income and Employment, Forestry and
Food Security. The Oxford Forestry Institute. Satellite Paper.

Ben Salem, B. Sand Dune Ecology and Rehabilitation: Implications
for Food Security. Satellite Paper.

Brokensha, D. and Castro, A.H.P. Common Property Resources. The
Institute for Development Anthropology. Satellite Paper.

Castro, A.H.P. and Brokensha, D. Institutions and Food Security:
Implications for Forestry Development. The Institute for
Development Anthropology. Main Paper.

Castro, A.H.P. and Brokensha, D. Landholding Systems and Agrarian
Change. The Institute for Development Anthropology. Satellite
Paper.

Child, B. and Child, G. The Effect of Institutions on the Economic
Price and Use of Wildlife. Satellite Paper.

Falconer, J. Forestry and Diets. The Oxford Forestry Institute.
Satellite Paper.

Hamilton, L.S. The Environmental Influences of Forests and Forestry
in Enhancing Food Production and Food Security. The East West
Center. Main Paper.

Nair, P.K.R. Institutional and Policy Aspects of Agroforestry. The
International Council for Research in Agroforestry (ICRAF).
Satellite Paper.

Nair, P.K.R. Production Systems and Production Aspects. The
International Council for Research in Agrforestry (ICRAF). Main
Paper.

Niamir, M. Forestry and Food Security. Synthesis Paper.

Oxby, C. Social Change in Forest Communitites Associated with
Recent Developments in Forest Area. The Oxford Forestry Institute.
Satellite Paper.

Rollet, B. Contributions of Mangrove Ecosystems in Food Production
and Security. Satellite Paper.

Seal, K. and Le Lucas, G. Conservation of Plant Genetic Resources
and Its Role in Improved Food Production and Security. The Royal
Botanic Gardens, Kew. Satellite Paper.

Swift, J. and Purata, S.E. Forestry and Food Security in the
Pastoral Economies of Northern Tropical Africa. The Institute of
Development Studies, University of Sussex. Satellite Paper.

Upadhyay, K.P. Forestry and Food Security in the Densely Populated
Upland Ecosystems of the Himalayas. Satellite Paper.

Venero, J.L.G. Forestry and Secure Food Production in the Peruvian
Andes. Satellite Paper.

Bibliography

Agarwal, B. Cold Hearths and Barren Slopes, The Woodfuel Crisis
1986   in the Third World. Zed Books Ltd., London.

Agarwal, B. and Deshingkar, P. Headloaders: Hunger for
1983   Fuelwood-I. CSE Reports 1983, No. 118. Centre for
          Science and Environment, Delhi.

Ajayi, S.S. Utilization of Forest Wildlife in West Africa. Report
1979   prepared for the Forestry Department, FAO, Rome.
          Misc/79/26.

Alcantara, E. et al. Crisis de Energia Rural y Trabajo Feminino
1982   en Tres Areas Ecologicas del Peru. World Employment
          Programme Research Working Papers 10, ILO, Geneva.

Ardayfio, E. Energy and Rural Women's Work in Ghana. in Energy
1985   and Rural Women's Work, Vol. II. Papers of a
          Preparatory Meeting, ILO, Geneva.

Arnold, J.E.M. et al. Evaluation of the SIDA Supported Social
1988   Forestry Project in Tamil Nadu, India. SIDA Evaluation
          Report No. 8, Stockholm.

Asibey, E.O.A. Wildlife and Food Security. Paper prepared for
1987   FAO Forestry Department, Rome.

Bach, W. The Potential Consequences of Increasing CO2 Levels in
1978   the Atmosphere. In Williams, J. (ed.) Carbon Dioxide,
          Climate and Society. Pergamon Press, Oxford. Pp.
          141-167.

Becker, B. The Contribution of Wild Plants to Human Nutrition
1983   in the Ferlo, Northern Senegal. Agroforestry Systems
          1:257-267.

Becker, B. Wild Plants for Human Nutrition in the Sahelian
1986   Zone. Journal of Arid Environments 11(1) 61-64.

Beer, J. Advantages, Disadvantages and Desirable
1987   Characteristics of Shade Trees for Coffee, Cacao and
          Tea. Agroforestry Systems 5 (1) 3-13.

Bell, T.I.W. Erosion in the Trinidad Teak Plantations.
1973   Commonwealth Forestry Review 52:223-233.

Bernard, E.A. L' Evapotranspiration Annuelle de la Foret
1953   Equatoriale Congolaise et l'Influence de celle-ci sur
          la Pluviosite. Pp. 201-204 in Proceedings of the IUFRO
          (International Union of Forest Research Organizations)
          Congress, Rome.

Bhimaya, C.P. Shelterbelts - Functions and Uses. In Conservation
1976   in Arid and 1976 Semi-Arid Zones. Conservation Guide
          3. FAO, Rome. Pp. 17-28.

Bille, J.C. Etude de la Production Primaire Nette d'un
1977   Ecosysteme Sahelien. Traveaux et Documents de l'ORSTOM
          No. 65. Paris.

Bishop, J.P. Tropical Forest Sheep on Legume Forage/Fuelwood
1983   Fallows. Agroforestry Systems 1(2):79-84.

Blackie, J.R. Hydrologic Effects of a Change in Land Use from
1972   Rain Forest to Tea Plantation in Kenya. In Studies and
          Reports in Hydrology No. 12, International Association
          of Hydrological Sciences, UNESCO. Pp. 312-329.

Bogneteau-Verlinden, E. Study on the Impacts of Windbreaks in
1980   the Majjia Valley, Niger. Agricultural University,
          Wageningen, and Cooperative for American Relief
          Everywhere, New York.

Bonkoungou, E.G. Acacia albida - a Multipurpose Tree for Arid
1985   and Semi-arid Zones. Forest Genetic Resources
          Information, FAO, No. 13:30-36.

Boomgard, J.J. The Economics of Small Scale Furniture Production
1983   and Distribution in Thailand. Ph. D. Dissertation,
          Michigan State University, Department of Agricultural
          Economics, Lansing.

Bormann, F.H. and Likens, G.E. Pattern and Process in a Forested
1981   Ecosystem. Springer-Verlag, New York.

Bosch, J.M. and Hewlett, J. D. A Review of Catchment Experiments
1982   to Determine the Effect of Vegetation Changes on Water
          Yield and Evapotranspiration. Journal of Hydrology
          55(1):3-23.

Boudet, G. and Toutain, B. The Integration of Browse Plants
1986   within Pastoral and Agro-pastoral Systems in Africa.
          In le Houerou, H. (ed), Browse in Africa, ILCA, Addis
          Ababa. Pp. 427-432.

Boughton, W.C. Effects of Land Management on Quantity and
1970   Quality of Availability Water: A Review. Australian
          Water Resources Council Research Project 68/2, Report
          120. University of New South Wales, Manly Vale.

Brooks, K.N. Evaluation of Deforestation: Impacts on Environment
1985   and Productivity. In Proceedings of the Ninth World
          Forestry Congress, Mexico. Section E-1.6.1.A.

Bruijnzeel, L.A. Hydrological and Biochemical Aspects of Manmade
1983   Forests in South-Central Java, Indonesia. Free
          University of Amsterdam, Amsterdam.

Bruijnzeel, P.S. Environmental Impacts of (De)forestation in the
1986   Humid Tropics: A Watershed Perspective. Wallaceana
          46:3-13.

Buch, A.N. and Bhatt, E.R. The Economic Status of Women Firewood
1980   Pickers from Mt. Girnar Junagadh, Gujarat. In Mathur
          et al (eds.) Proceedings of the Seminar on the Role of
          Women in Community Forestry, Dehra Dun, India, Dec.
          4-9, 1980. Indian Forest Service.

Butynski, T.M. and von Richter, W. In Botswana Most of the Meat
1974   is Wild. Unasylva 26(106):24-29.

Caborn, J.M. Shelterbelts and Windbreaks. Faber and Faber,
1965   London.

Campbell, A. The Use of Wild Food Plants in Drought in Botswana.
1986b   Journal of Arid Environments 11(1):81-91.

Campbell, B.M. The Importance of Wild Fruits for Peasant
1986a   Households in Zimbabwe. Food and Nutrition
          12(1):38-44.

Campbell-Platt, G. African Locust Bean (Parkia sp.) and its
1980   Fermented Food Product Dawadawa. Ecology of Food and
          Nutrition 9(2):123-132.

Castro, A. et al. Indicators of Rural Inequality. World
1981   Development 9(5): 401-427.

Castro, A. Household Energy Use and Tree Planting in
1983   Kirinyaga. University of Nairobi, Institute for
          Development Studies Working Paper, Nairobi.

Cecelski, E. The Rural Energy Crisis Women's Work and Basic
1984   Needs: Perspectives and Approaches to Action. World
          Employment Programme Research Working Paper, ILO,
          Geneva.

Cernea, M. Alternative Units of Social Organization for
1985   Sustaining Afforestation Strategies. In Putting People
          First: Sociological Variables in Rural Development.
          Cernea, M. (ed.), Oxford University Press, Oxford.

Chambers, R. Rural Development: Putting the Last First. Longman,
1983   New York.

Chambers, R. and Longhurst, R. Trees, Seasons, and the Poor. IDS
1986   Bulletin 17(3):4450. Institute of Development Studies,
          University of Sussex, Sussex.

Chambers, R.E. Albedo of Tropical Crops: Implications for Water
1980   Balance Studies. In Furtado, J. (ed), Tropical Ecology
          and Development: Proceedings of the Fifth
          International Symposium of Tropical Ecology, Kuala
          Lumpur. Pp. 549-552.

Chambers, R. and Leach, M. Trees to Meet Contingencies: Savings
1987   and Security for the Rural Poor. Discussion Paper
          228. Institute of Development Studies, University of
          Sussex, Sussex.

Chepil, W.S. Dynamics of Wind Erosion. Soil Science
1945   60(5):397-411.

Christensen, B. Mangroves: What are they worth? Unasylva
1983   35(139):2-15.

Connelly, W. Copal and Rattan Collecting in the Philippines.
1985   Economic Botany 39(1):39-46.

Conway, F.J. Case Study: The Agroforestry Outreach Project in
1987   Haiti. Paper presented at the IIED Conference on
          Sustainable Development, International Institute for
          Environment and Development, London.

Costen, E.   Arid Zone Examples of Shelterbelt Establishment and
1976   Management. In Conservation in Arid and Semi-Arid
          Zones. FAO Conservation Guide 3, Rome. Pp. 29-41.

Daly, J.J. Cattle Need Shade Trees. Queensland Agricultural
1984   Journal 110(1):21-24.

Dani, Anis A. et al, Institutional Development for Local
1987   Management of Rural Resources. East-West Environment
          and Policy Institute, Honolulu.

Deveau, L.E. and Castle, J.R. The Industrial Development of
1976   Farmed Marine Algae: The Case History of Eucheuma in
          the Philippines and U.S.A. In FAO Technical Conference
          on Aquaculture, Kyoto, Japan, 26 May 1976. FAO Fishery
          Resources and Environment Division, Rome.

Dirar, H.A. Kawal, Meat Substitutes from Fermented Cassia
1984   obtusifolia leaves. Economic Botany 38(3): 342-349.

Dommergues, Y.R. The Role of Biological Nitrogen Fixation in
1987   Agroforestry. In Steppler and Nair (eds.),
          Agroforestry: A Decade of Development. ICRAF, Nairobi.

Douglass, J.E. A Summary of Some Results from the Coweeta
1983   Hydrologic Laboratory. In Hamilton and King (eds.),
          Tropical Forested Watersheds: Hydrologic and Soil
          Responses to Major Uses and Conversions. Westview
          Press, Boulder. Pp. 137-141.

Dourojeanni, M.J. The Integrated Management of Forest Wildlife
1978   as a Source of Protein for Rural Populations. Paper
          presented at the Eighth World Forestry Congress,
          Jakarta. Agenda Item No. B. FFF/8-0.

Dyson-Hudson, N. Pastoral Production Systems and Livestock
1985   Development Projects: An East African Perspective. In
          Cernea, M. (ed.), Putting People First: Sociological
          Variables in Rural Development. Oxford University
          Press, Oxford.

Eardley-Wilmot, S. Notes on the Influence of Forests on the
1906   Storage and Regulation of Water Supply. Forests
          Bulletin No. 9. Govt. Printing Press, Calcutta.

Ekern, P.C. Direct Interception of Cloud Water on Lanaihale,
1964   Hawaii. Proceedings Soil Science Society of America
          28(3):419-421.

Engel, A. et al. Promoting Smallholder Cropping Systems in
1985   Sierra Leone. An Assessment of Traditional Cropping
          Systems and Recommendations for the Bo-Pujehun Rural
          Development   Project.   Studien   No.IV/86.
          Schriftenreihedes   Fachbereichs   Internationale
          Agarentwicklung, Technische Universitat, Berlin.

Erasmus, C.J. In Search of the Common Good. Free Press, New
1977   York.

Falconer, J. Household Food Security and Forestry: An Analysis
1989a   of Socio-Economic Issues. FAO, Rome.

Falconer, J. The Major Significance of 'Minor Forest Products':
1989b   Local People's Uses and Values of Forests in the Humid
          Forest Region of West Africa. FAO, Rome.

FAO.   Guidelines   for   Land   Evaluaticn   for   Rainfed
1977   Agriculture. FAO Soils Bulletin 52. FAO, Rome.

FAO.   Fruit-bearing Forest Trees. FAO Forestry Paper No. 34,
1982   Rome.

FAO.   India, Malaysia and Thailand. A Study of Forest as a
1983a   Source of Food. FAO, Bangkok.

FAO.   Food and Fruit-bearing Forest Species. Examples from
1983b   East Africa. FAO Forestry Paper No. 44:1, Rome.

FAO.   Food and Fruit-bearing Forest Species. Examples from
1984   South East Asia. FAO Forestry Paper No. 44:2, Rome

FAO.   Sand   Dune   Stabilization,   Shelterbelts   and
1985   Afforestation in Dry Zones. FAO Conservation Guide 10,
            Rome.

FAO.   Some Medicinal Forest Plants of Africa and Latin
1986a   America. FAO Forestry Paper No. 67, Rome.

FAO.   Food and Fruit-bearing Forest Species. Examples from
1986b   Latin America. FAO Forestry Paper No. 44:3, Rome.

FAO.   Small-scale Forest-Based Enterprises. Forestry Paper
1987   No. 79, Rome.

Felker, P. Mesquite, an All-purpose Leguminous Arid Land Tree.
1979   In Ritchie (ed.), New Agricultural Crops. Westview
            Press, Boulder.

Felker, P. and Bandurski, R.S. Uses and potential uses of
1979   leguminous trees for minimal energy input agriculture.
            Economic Botany 33(2): 172-184.

Fernandes, E.C.M. and Nair, P.K. An evaluation of the structure
1986   and function of some tropical home gardens.
            Agricultural Systems 21: 179-310.

Fisseha, Y. Basic Features of Rural Small-scale Forest-based
1987   Processing Enterprises. In Small-scale Forest-based
            Processing Enterprises. FAO Forestry Paper No. 79,
            Rome.

Fleuret, A. The Role of Wild Foliage in the Diet: A Case Study
1979   from Lushoto, Tanzania. Ecology of Food and Nutrition
            8(2):87-93.

Fortmann, L. The Tree Tenure Factor in Agroforestry with
1984   Particular Reference to Africa. Agroforestry Systems
            2:231-248.

Fortmann, L. and Riddell, J. Trees and Tenure: An Annotated
1984   Bibliography for Agroforesters and Others.   ICRAF,
            Nairobi.

van Gelder, B. and Kerkhof, P. The Agroforestry Survey in
1984   Kakamega District: Final Report. Working Paper No.
            6, Kenya Woodfuel Development Programme, Beijer
            Institute, Nairobi.

Gielen, H. Report on an Agroforestry Survey in Three Villages
1982   of North Machakos, Kenya. ICRAF, Nairobi.

Giffard, P.L. Les gommiers, essences de reboisement pour les
1975   regions saheliens. Bois et Forets des Tropiques
            161:321.

Gilmour, D.A. The Effects of Logging on Streamflow and
1971   Sedimentation in a North Queensland Rainforest
            Catchment. Commonwealth Forestry Review 50(1):39-48.

Glover, N. and Beer, J. Nutrient Cycling in Two Traditional
1986   Central American Agroforestry Systems. Agroforestry
            Systems 4(2): 77-87.

Goodland, R.J.A. and Irwin, H.S. Amazon Jungle: Green Hell to
1975   Red Desert? Elsevier Press, Amsterdam.

Gorse, J.   Desertification in the Sahelian and Sudanian Zones
1985   of West Africa. Unasylva 150, 37(4):2-18.

Grandstaff, S.W., et al. Trees in Paddy Fields in Northeast
1986   Thailand.   In   Marten, G. (ed.), Traditional
            Agriculture in Southeast Asia. Westview Press,
            Boulder. Pp. 273-292.

Grivetti, L.E. Dietary Resources and Social Aspects of Food Use
1976   in a Tswana Tribe. PhD Dissertation. Department of
            Geography, University of California at Davis.

Grossman, L. Peasants, Subsistence, Ecology and Development in
1984   the Highlands of Papua New Guinea. University Press.

Guyot, G. Brise-vent et rideaux abris avec reference
1986   particuliere aux zones seches. FAO Conservation Guide
          No. 15, Rome.

Hamilton, L.S. Tropical Rainforest Use and Preservation: A Study
1976   of Problems and Practices in Venezuela. International
          Series No. 4, Sierra Club, San Fransisco.

Hamilton, L.S. (with King, P.N.) Tropical Forested Watersheds:
1983   Hydrologic and Soils Response to Major Uses or
          Conversions. Westview Press, Boulder.

Hamilton, L.S. Overcoming Myths about Soil and Water Impacts of
1986   Tropical Forest Land Uses. In El Swaify, S.A. et al
          (eds.) Soil Erosion and Conservation. Soil
          Conservation Society of America, Ankeny. Pp. 680-690.

Hamilton, L.S. The Environmental Influences of Forests and
1988   Forestry in Enhancing Food Production and Food
          Security. The East West Center. Main Paper presented
          at the FAO Expert Consultation on Forestry and Food
          Production/Security, India, 8-20 Feb. 1988.

Hamilton, L.S. and Pearce, A.J. Biophysical Aspects of
1986   Integrated Watershed Management. In Easter, K.W. et
          al (eds.), Watershed Management: An Interdisciplinary
          Approach. Studies in Water Policy and Management No.
          2:10. Westview Press, Boulder. Pp. 33-52.

Hamilton, L.S. and Snedaker, S.C. (eds.) Handbook for Mangrove
1984   Area Management. Environment and Policy Institute,
          East West Center, IUCN/UNESCO/UNEP, Honolulu.

Hammer, T. Reforestation and Community Development in Sudan.
1982   Resources for the Future, Washington D.C.

Hampicke, U. Man's Impact on the Earth's Vegetation Cover and
1979   its Effects on Carbon Cycle and Climate. In Bach, W.
          et al (eds.) Man's Impact on Climate. Elsevier Press,
          Amsterdam. Pp. 139-159.

Hardin, G. The Tragedy of the Commons. Science 162: 1243-1248
1968   (December 13).

Hardjono, H.W. The Effect of Permanent Vegetation and its
1980   Distribution on Streamflow of Three Sub-watersheds in
          Central Java. Paper presented at Seminar on Hydrology
          and Watershed Management, Surakarta (5 June).

Hassan, N. et al. Seasonal Patterns of Food Intake in Rural
1985   Bangladesh: Its Impact on Nutritional Status. Ecology
          of Food and Nutrition 17(2):175-186.

Havnevick, K. Analysis of rural production and income, Rufiji
1980   Districts, Tanzania. CHR Michelsen Institute, DERAP
          Publication No. 152,137. Jointly issued with the
          Institute of Resource Assessment, University of Dar
          es Salaam. Research Paper No. 3.

Heinz, H.J. and Maguire, B. The Ethnobiology of the Iko Bushmen:
1974   Their Ethnobotanical Knowledge and Plant Lore.
          Occasional Paper No. 1, Botswana Society, Gabarone.

Hendersen-Sellers, A. and Gornitz, V. Possible Climatic Impacts
1984   of Land Cover Transformations. Climatic Change
          6:231-257.

Herring, R. Land to the Tiller. Yale University Press, New
1983   Haven.

Hewlett, J.D. Forests and Floods in the Light of Recent
1982   Investigation. In Proceedings of the Canadian
          Hydrological Symposium: Hydrologic Processes of
          Forested Areas. Fredericton, N.B. National Research
          Council, Ottawa. Pp. 543-560.

Hill, M. The Relation between Forests and Atmospheric and
1916   Soil Moisture in India. Forestry Bulletin (New
          Series, Calcutta) No. 33.

Holmes, J.W. and Wronski, E.B. On the Water Harvest from
1982   Afforested Catchments. In E.M. O'Loughlin and L.J.
          Bren (eds.), The First National Symposium on Forest
          Hydrology. Institute of Engineers, Barton. Pp. 1-6.

le Houerou, H.N. (ed.) Browse in Africa: The Current State of
1986   Knowledge. IBPGR and Royal Botanic Gardens, Kew,
          London.

Hough, J.   Management Alternatives for Increasing Dry Season
1986   Base Flow in the Miombo Woodlands of Southern Africa.
           Ambio 15(6):341-346.

Hughes, K.K. Trees and Salinity. Queensland Agricultural Journal
1984   110 (1):13-14.

Hussain, M.A. Seasonal Variation and Nutrition in Developing
1985   Countries. Food and Nutrition 11(2):23-27.

IDRC.   Rattan: A Report of a Workshop Held in Singapore.
1980   Ottawa, Canada.

Irvine, F.R. Supplementary and Emergency Food Plants of West
1952   Africa. Economic Botany 6(1):23-40.

Jackson, J.K. and Boulanger. The Forest of the Mae Sa Valley
1978   Northern Thailand as a Source of Food. 8th World
           Forestry Congress, Jakarta. Vol. III, pp. 1059-1063.

Jacobs, A. Pastoral Maasai and Tropical Rural Development. In
1980   R. Bates and M. Lofchie (eds.), Agricultural
           Development in Tropical Africa, Praeger, New York.

Jahn, S.A.A.A. et al. The Tree that Purifies Water: Cultivating
1986   Multipurpose Moringaceae in the Sudan. Unasylva
           38(152):23.

Janzen, D.H. Additional Land at What Price? Responsible Use
1976   of the Tropics in a Food-Population Confrontation.
           Proceedings American Phytopathological Society
           3:35-39.

Jensen, A.M. Les Effets des Brise-vent en Zones Temperee et
1984   Tropical. Manuscript Report IDRC-MR800. International
           Development Research Center, Ottawa.

Jones, R.J. The Value of Leucaena leucocephala as a Feed for
1979   Ruminants in the Tropics. World Animal Review 31:
           13-23.

Juo, A.S.R. and Lal, R. The Effect of Fallow and Continuous
1977   Cultivation on the Chemical and Physical Properties of
           an Alfisol in Western Nigeria. Plant and Soil 47(3):
           567-584.

Kamara, J.N. Firewood Energy in Sierra Leone: Production,
1986   Marketing and Household Use Patterns. Studien zur
           Integrierten Landlichen Entwicklung Verlag Weltarchiv
           No. 9. Hamburg.

Kammer, R. and Raj. Preliminary Estimates of Minimum Flows in
1979   Varaciva Creek and the Effect of Afforestation on
           Water Resources. Fiji Public Works Department
           Technical Note 7911, Suva.

Kang, B.T., Gromme, H. and Lawson, T. Alley Cropping
1985   Sequentially Cropped Maize and Cowpea with Leucaena on
           a Sandy Soil in Southern Nigeria. Plant and Soil 85:
           267-277.

Kang, B.T. and Lal, R. Nutrient Losses in Water Runoff from
1981   Agricultural Catchments. In R. Lal and Russell (eds.),
           Tropical Agricultural Hydrology. John Wiley and Sons,
           New York. Pp. 153-61.

Kapetsky, J.M. The Mangrove Ecosystem for Forestry, Fisheries
1987   and Aquaculture. Symposium on Ecosystem
           Redevelopment: Ecological, Economic and Social
           Aspects, Budapest, 5-10 April.

Karg, J. Influence of Shelterbelts on Distribution and
1976   Mortality of Colorado Beetle (Leptinotarsa
           decemlineata Say). In Table Ronde CNRS: Aspects
           Physiques, Biologiques et Humains des Ecosystemes
           Bocagers des Regions Temperees Humides. Les
           Bocages-Histoire, Ecologie, Economie. INRA, ENSA et
           Universite de Rennes, Rennes.

Kilby, P. and Liedholm, C. The Role of Non-farm Activities in
1986   the Rural Economy. EEPA Discussion Paper No. 7.

de Kock, G.C. Drought Resistant Fodder Crops. Proceedings of
1967   Grasslands Seminar on Africa, 2:147-156.

Kostantinov, A.R. and L.R. Struzer. Shelterbelts and Crop
1965   Yields. Gidrometeorologicheskor Izdalel Stvo,
           Leningrad. (Translation from Russian by Israel Program
           for Scientific Translations, 1969.)

Krishnamurthy, K. Humans' Impact on the Pichavaram Mangrove
1984   Ecosystem: A Case Study from Southern India.
            Proceedings of the Asian Symposium on Mangrove
            Environment, Research and Management, Kuala Lumpur 25-
            29 August 1980. Pp. 624-632.

Kuchar, P. The Status and Significance of Yicib - _Cordeauxia_
1986   _edulis_ - in the Central Rangelands of Somalia. _In_
            Proceedings of the Seminar of Future Development
            Strategies for Range/Livestock Development in
            Central Somalia.

Kulkarni, D.H. and Junagad, C.F. Utilization of Mangrove Forests
1959   in Saurashtra and Kutch. _In_ Proceedings of the
            Mangrove Symposium, Calcutta, 16-19 Oct. 1957. Govt.
            India Press, Faridabad. Pp. 30-35.

Lagemann, J. Traditional African Farming Systems in Eastern
1977   Nigeria: An Analysis of Reaction to Increasing
            Population Pressure.   Africa Studien, Weltforum
            Verlag, Munich.

Lai, Chun and Khan, A. Mali as a Case Study of Forest Policy in
1986   the Sahel: Institutional Constraints on Social
            Forestry. Social Forestry Network Paper, No. 3e.
            Overseas Development Institute, London.

Lal, R. and Cummings, D.J. Clearing a Tropical Forest I. Effects
1979   on Soil and Microclimate. Field Crops Research
            2(2):91-107.

Langford, K.J. Change in Yield of Water Following a Bushfire in
1976   a Forest of _Eucalyptus regnans_. Journal of Hydrology
            (Netherlands) 29(1/2): 87-114.

Lembaga Ekologi. Report on Study of Vegetation and Erosion in
1980   the Jatiluhur Catchment. Institute of Ecology,
            Bandung.

Lettau, H., Lettau, K. and Molion, L.C. Amazonia's Hydrologic
1979   Cycle and the Role of Atmospheric Recycling in
            Assessing Deforestation Effects. Monthly Weather
            Review 107:227-238.

Longhurst, R. Cropping Systems and Household Food Security:
1985   Evidence From Three West African Countries. Food and
            Nutrition 121(2):10-16.

Longhurst, R. Cash Crops, Household Food Security and Nutriton.
1987   Cash Crops Workshop, Institute of Development Studies
            at the University of Sussex, Sussex.

Lusigi, W.J. Combatting Desertification and Rehabilitating
1981   Degraded Production Systems in Northern Kenya.
            UNESCO-UNEP.   Integrated Project in Arid Lands
            Technical Report No. A-4. UNESCO, Nairobi.

Malaisse,F. and Parent, G. Edible Wild Vegetable Products in the
1985   Zambezian Woodland Area: A Nutritional and Ecological
            Approach. Ecology of Food and Nutrition 18:43-82.

Masson, J.L. Integrated Development of the Sundarbans Forest
1984   Reserve, Bangladesh. Technical Report TCP/BGD/2309,
            FAO, Rome.

Mathur, H.N., Babu, R., Joshie, P. and Singh, B. Effect of
1976   Clearfelling and Reforestation on Runoff and Peak
            Rates in Small Watersheds. Indian Forester
            102(4):219-26.

May, P.H. _et al_. Babassu Palm in the Agroforestry Systems in
1985a   Brazil's Mid-north Region. Agroforestry Systems
            3(39):275-295.

May, P.H. _et al_. Subsistance Benefits from Babassu Palm.
1985b   Economic Botany 39(2):113-129.

Mead, D.C.   Small Industries in Egypt: An Exploration of the
1982   Economics of Small Furniture Producers. International
            Journal of Middle East Studies 14:159-171.

Megahan, W.F. and King, P.N. Identification of Critical Areas on
1985   Forest Lands for Control of Non-point Sources of
            Pollution. Environmental Management 9(1):7-17.

Michon, G., Bompard, J., Hecketsweiler, P. and Ducatillion, C.
1983   Tropical Forest Architectural Analysis as Applied to
            Agroforests in Humid Tropics: The Example of
            Traditional Village-agroforests in West Java.
            Agroforestry Systems 1(2):117-129.

Mnzava, E.M. Village Industries and Savannah Forests. Unasylva
1981   33(131):24-29.

Mungkorndin, S. Forests as a Source of Food to Rural
1981   Communities in Thailand. FAO, NO. RAPA 52, Bangkok.

Murray, G.   Mountain Peasants in Honduras: Guidelines for the
1981   Reordering of Smallholding Adaptation to the Pine
          Forest. USAID, Tegucigalpa.

Myers, N.   Conversion of Tropical Moist Forests. National
1980   Research Council, Washington.

Nair, P.K.R. Soil Productivity Aspects of Agroforestry. Science
1984a   and Practice of Agroforestry 1. International Council
          for Research in Agroforestry (ICRAF), Nairobi.

Nair, P.K.R. Fruit Trees in Agroforestry. Working Paper.
1984b   Environment and Policy Institute, East-West Center,
          Honolulu.

Nair, P.K.R. Agroforestry Systems Inventory. Agroforestry
1987a   Systems 5: 301-317.

Nair, P.K.R. Use of Perennial Legumes in Asian Farming Systems.
1987b   Paper presented to the International Workshop on
          Sustainable Agriculture, IRRI, The Philippines.

Nair, P.K.R. Production Systems and Production Aspects. The
1988   International Council for Research in Agroforestry
          (ICRAF). Main Paper preseneted at the FAO Expert
          Concultation on Forestry and Food Production/Security,
          India, 8-20 Feb.

National Academy of Sciences (NAS). Firewood Crops.
1980   National Academy of Sciences (NAS), Washington D.C.

National Academy of Sciences (NAS). Firewood crops. Vol. 2.
1983   National Academy of Sciences (NAS), Washington D.C.

O'Loughlin, C.L. and Watson, A.J. Root Wood Strength
1981   Deterioration in Beech (<u>Nothofagus fusca</u> and <u>N.
          truncata</u>) after Clearfelling. New Zealand Journal of
          Forestry Science 11(2):183-185.

Ogle, B.M. and Grivetti, L.E. Legacy of the Chameleon: Edible
1985   Wild Plants in the Kingdom of Swaziland, South Africa.
          A Cultural, Ecological, Nutritional Study. Ecology of
          Food and Nutrition, Part I 16(3):193-208, Parts II-IV.
          17(1):1-60.

Olofson, H. Indigenous Agroforestry Systems. Philippine
1983   Quarterly of Culture and Society 11: 149-174.

Parent, G.   Food Value of Edible Mushrooms from Upper Shaba.
1977   Economic Botany 31:436-445.

Parkinson, S. Nutrition in the South Pacific: Past and Present.
1982   Food and Nutrition 39(3):121-125.

Pearce, A.J. Erosion and Sedimentation. Environment and Policy
1986   Institute Working Paper. East-West Center, Honolulu.

Penny, D.H. and Singarimbun, T. Population and Poverty in Rural
1973   Java: Some Economic Arithmetic from Sriharjo. Cornell
          International Agricultural Development Monograph 41,
          Department   of   Agricultural   Economics,   Cornell
          University, Ithaca.

Persson, R. World Forest Resources. Royal College of Forestry,
1974   Stockholm.

Potter, G.L., Ellsaesser, H.W., MacCraken, M.C. and Luther, F.M.
1975   Possible Climatic Impact of Tropical Deforestation.
          Nature 258(5537):697-698.

Poulsen, G. Man and Trees in Tropical Africa. International
1978   Development Research Center (IDRC), Research Report
          No. 101C, Ottawa.

Poulsen, G. The Non-Wood Products of African Forests. Unasylva
1982   34(137):15-21.

Prachett, D., <u>et al.</u> Factors Limiting Liveweight Gain of Beef
1977   Cattle on Rangeland in Botswana. Journal of Range
          Management 30: 442-445.

Raintree, J.B. and Warner, K. Agroforestry Pathways for the
1986   Intensification of Shifting Cultivation. Agroforestry
          Systems 4(1): 39-54.

Rambo, A.T.   Why Shifting Cultivators Keep Shifting:
1984   Understanding Farmer Decision-making in Traditional
             Agroforestry Systems. In Community Forestry: Some
             Aspects. UNDP/East-West Center/FAO, Bangkok.

Ranganathan, C.R. Protective Functions of Forests. Paper
1949   Presented at the UN Scientific Conference on
             Resources, Lake Success (Also in the Indian Forester
             76(1):2-11.)

Reinhart, K.G., Eschner, A.R. and Trimble, G.R. Jr. Effects on
1963   Streamflow of Four Forest Practices in the Mountains
             of West Virginia. U.S. Forest Service Research Paper
             No. NE1, Northeastern Forest Experiment Station, Upper
             Darby.

Runge, C.   Common Property and Collective Action in Economic
1986   Development. World Development 14(5):623-635.

Saenger, P. et al. Global Status of Mangrove Ecosystems.
1983   Commission on Ecology Paper No 3, IUCN, Gland.

Salati, E. and Vose, P.B. Amazon Basin: A System in Equilibium.
1984   Science 225(4658):129-138.

Sanchez, P.A. Soil Productivity and Sustainability in
1987   Agroforestry Systems. In H.A. Steppler and P.K.R. Nair
             (eds.), Agroforestry: A Decade of Development. ICRAF,
             Nairobi.

Sedjo, R.A. and Clawson, M. Global Forests. In J. Simon and H.
1984   Kahn (eds.), The Resourceful Earth: A Response to
             Global 2000. Basil Blackwell, Oxford. Pp. 128-170.

Shpak, I.S. Effect of Forest on Water Balance Components of
1968   Drainage Basins. Academy of Sciences of the Ukraine
             USSR, Kiev. Pp. 137-143. (Israel Program for
             Scientific Translations, 1971. United States
             Department of Agriculture, Washington.)

Siccama, T.G. and Smith, W.H. Lead Accumulation in a Northern
1978   Hardwood Forest. Environmental Science and Technology
             12(5):593-594.

Siebert, S.F. and Belsky, J.M. Forest Product Trade in a Lowland
1985   Filipino Village. Economic Botany 39(4): 522-533.

Singh, R.V. Forestry and Food Security in India. Indian
1988   Council of Forestry Research and Education, Dehra Dun.

Skerman, P.J. Legiminous Browse: Tropical Forage Legumes. FAO
1977   Plant Production and Protection Series No.2, Rome. Pp.
             431-525.

Skutsch, M. The Not-So-Social Forestry in Gujarat, India. VOK
1987   Working Paper No. 29, Technology and Development
             Group, University of Twente, Twente.

Soemarwoto, O. and Soemarwoto, I. The Javanese Rural Ecosystem.
1984   In Rambo, A.T. and P.E. Sajise (eds), An Introduction
             to Human Ecology Research on Agricultural Systems in
             Southeast Asia. University of the Philippines, Los
             Banos.

Spears, J.   Rehabilitating Watersheds. Finance and Development
1982   19(1):30-33.

Stoler, A.   Garden Use and Household Consumption Pattern in a
1975   Javanese Village. Ph. D. Thesis,   Department of
             Anthropology, Columbia University, New York.

Stoler, A.   Garden Use and Household Economy in Rural Java.
1978   Journal of Indonesian Studies 14(2): 85-101.

Swaminathan, M.S. Can Africa Feed Itself? An Application of
1986   Lessons Learned in Asia to the Challenge Facing
             Africa. The Hunger Project, First Annual Tanco
             Memorial Lecture. World Food Council, Rome.

Swanson, F.J., Swanson, M.M., and Woods, C. Analysis of Debris
1981   Avalanche Erosion in Steep Forested: An Example from
             Mapleton, Oregon, U.S.A. International Association of
             Scientific Hydrology Publication 132:67-75.

Terra, G.T.A. Mixed-garden Horticulture in Java. Malaysian
1954   Journal of Tropical Geography 4: 33-43.

Torres, F.   The Role of Woody Perennials in Animal
1983   Agroforestry. Agroforestry Systems 1(2): 131-163.

Trollope, W.S.W. The Growth of Shrubs and Trees and their
1981   Reaction to Treatment. In N.M. Tainton (ed.), Veld and
            Pasture Management in South Africa. (Pietermaritzburg,
            Shuter & Shooter, South Africa.)

TropSoils. Annual Report for 1985-86. Soil Science Department,
1986   North Carolina University, Raleigh.

Tushaar Shah, Gains from Social Forestry: Lessons from West
1987   Bengal. IDS/ODI Workshop on Commons, Wastelands,
            Trees and the Poor: Finding the Right Fit, University
            of Sussex, June. (Published as an ODI Social Forestry
            Network Paper.)

Umali, D.L. Problems of Shifting Cultivators and Forest
1987   Squatters and the Role of NGO's in National Forestry
            Programs, Statement at Asian Regional Workshop on
            Expanding the Role of NGO's in National Forestry
            Programs, Bangkok, February.

UN Educational, Scientific, and Cultural Organization. Working
1972   Group on the Influence of Man on the Hydrologic Cycle.
            Influence of Man on the Hydrologic Cycle: Guide to
            Policies for the Safe Development of Land and Water
            Resources. In Status and Trends of Research in
            Hydrology, Paris. Pp. 31-70.

USAID.   Windbreak and Shelterbelt Sechnology for Increasing
1987   Agricultural Production. Science and Technology
            Division, Forest and Natural Resources Department
            USAID, Washington. D.C.

Vergara, N.T. and Briones, N.D. (eds.) Agroforestry in the Humid
1987   Tropics: Its Protective and Ameliorative Roles to
            Enhance Productivity and Sustainability. East-West
            Center and College, Honolulu & Southeast Asian
            Regional Center for Graduate Study and Research in
            Agriculture, Laguna.

Vis, M.   Interception, Drop Size Distributions and Rainfall
1986   Kinetic Energy in Four Colombian Forest Ecosystems.
            Earth Surface Processes and Landforms 11:591-603.

Wang Shiji. A Brief Account of Agroforestry Development in the
1988   Plains of China. Chinese Academy of Forestry, Beijing.

Warren, W.D.M. A Study of Climate and Forests in the Ranchi
1974   Plateau Part 1. Indian Forester 100 (4):229-234.

Weber, F.R. (with Stoney, C.) Reforestation in Arid Lands.
1986   Volunteers in Technical Assistance, Arlington.

Weber, F.R. and Hoskins, M. Agroforestry in the Sahel.
1983   Department of Sociology, Virginia Polytechnic
            Institute, Blacksburg.

Weinstock, J.A. Rattan: A Complement to Swidden Agriculture.
1983   Economic Botany 37(1): 56-68.

Whitehead, F.H. Phenotypic Adaptation in Wind Exposed Plants.
1965   Scientific Horticulture 17:53-60.

Wickens, G. Alternative Uses of Browse Species. In le Houerou
1986   (ed.), Browse in Africa: The Current State of
            Knowledge. IBPGR and Royal Botanic Gardens, Kew,
            London. Pp. 155-185.

Wickens, G. et al. (eds.) Plants for Arid Lands. Royal Botanic
1985   Gardens, Kew, London.

Wiersum, K.F. Surface Erosion under Various Tropical
1984   Agroforestry Systems. In C. O'Loughlin and A. Pearce
            (eds.), Proceedings of the Symposium on Effects of
            Forest Land Use on Erosion and Slope Stability. EAPI,
            East-West Center, Honolulu. Pp. 231-239.

Wilson, A.D. The Digestibility and Voluntary Intake of the
1977   Leaves and Shrubs by Sheep and Goats. Australian
            Journal of Agricultural Research 28: 501-508.

Wilson, A.D. and Harrington, G.N. Nutritive Value of Australian
1986   Browse Plants. In H.N. Le Houerou (ed.), Browse in
            Africa: Current State of Knowledge. Kew Gardens,
            London.

Woodwell, G.M., et al. The Biota and the World Carbon Budget.
1978   Science 199:141-146.

World Bank, Economics Issues and Farm Forestry. Working Paper
1986   prepared for the Kenya Forestry Sector Study. World
            Bank, Washington D.C. (mimeo).

World Water. How Trees Combat Droughts and Floods. World Water
1981   4(10):18.